SURVEY
RESEARCH AND
SAMPLING

THE SAGE QUANTITATIVE RESEARCH KIT

Survey Research and Sampling by *Jan Eichhorn* is the 4th volume in *The SAGE Quantitative Research Kit*. This book can be used together with the other titles in the *Kit* as a comprehensive guide to the process of doing quantitative research, but is equally valuable on its own as a practical introduction to Survey Research.

Editors of The SAGE Quantitative Research Kit:

Malcolm Williams – *Cardiff University, UK*

Richard D. Wiggins – *UCL Social Research Institute, UK*

D. Betsy McCoach – *University of Connecticut, USA*

Founding editor:

The late W. Paul Vogt – *Illinois State University, USA*

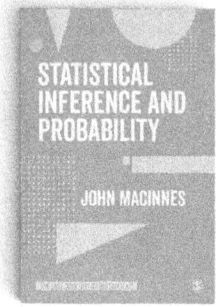

SURVEY RESEARCH AND SAMPLING

JAN EICHHORN

Los Angeles | London | New Delhi
Singapore | Washington DC | Melbourne

THE SAGE QUANTITATIVE RESEARCH KIT

Los Angeles | London | New Delhi
Singapore | Washington DC | Melbourne

SAGE Publications Ltd
1 Oliver's Yard
55 City Road
London EC1Y 1SP

SAGE Publications Inc.
2455 Teller Road
Thousand Oaks, California 91320

SAGE Publications India Pvt Ltd
B 1/I 1 Mohan Cooperative Industrial Area
Mathura Road
New Delhi 110 044

SAGE Publications Asia-Pacific Pte Ltd
3 Church Street
#10-04 Samsung Hub
Singapore 049483

Editor: Jai Seaman
Assistant editor: Charlotte Bush
Production editor: Manmeet Kaur Tura
Copyeditor: QuADS Prepress Pvt Ltd
Proofreader: Elaine Leek
Indexer: Cathryn Pritchard
Marketing manager: Susheel Gokarakonda
Cover design: Shaun Mercier
Typeset by: C&M Digitals (P) Ltd, Chennai, India

© Jan Eichhorn 2021

This volume published as part of *The SAGE Quantitative Research Kit* (2021), edited by Malcolm Williams, Richard D. Wiggins and D. Betsy McCoach.

Apart from any fair dealing for the purposes of research, private study, or criticism or review, as permitted under the Copyright, Designs and Patents Act, 1988, this publication may not be reproduced, stored or transmitted in any form, or by any means, without the prior permission in writing of the publisher, or in the case of reprographic reproduction, in accordance with the terms of licences issued by the Copyright Licensing Agency. Enquiries concerning reproduction outside those terms should be sent to the publisher.

Library of Congress Control Number: 2021931885

British Library Cataloguing in Publication data

A catalogue record for this book is available from the British Library

ISBN 978-1-5264-2380-1

CONTENTS

List of Tables and Boxes   ix
About the Author   xi

1   Introduction   **1**

Why Do We Need to Worry About Sampling and Survey Design?   2
The Approach of This Volume   3
Overview of the Content of This Volume   5

2   Sampling Approaches: How to Achieve Representativeness   **9**

Introduction   10
Populations and Sampling Frames   12
Probability and Non-Probability Sampling   15
Approaches to Probability Sampling   16
    Simple Random Sampling   16
    Cluster and Multistage Sampling   18
    Stratified Sampling   19
Approaches to Non-Probability Sampling   20
    Quota Sampling   21
    Common Alternative Non-Probability Sampling Methods   24
    Respondent-Driven Sampling   25

3   Sampling Mode: How We Actually Collect the Data   **31**

Introduction   32
Data Collection Methods   32
    Face-to-Face   33
    Telephone   35
    Direct Mail   37
    Internet Based   39
    Combining Methods   41
Considering Non-Response Bias   43
Adjustments After Data Collection: Weighting   46

	Applying Weights to Account for Response Biases	46
	Applying Weights to Account for Attrition	48
4	**Questionnaire Design: Asking the Right Questions**	**53**
	Introduction	54
	Distinguishing Between Different Types of Questions	55
	Open- and Closed-Ended Questions	55
	Single-Answer Multiple-Choice Questions	55
	Multiple-Answer-Item Questions	57
	Likert Scale Questions	58
	Numerical Scale Questions	60
	Ranking Questions	61
	Open Format (With Predetermined Units)	62
	What Makes Good Questions – and Some Common Pitfalls	62
	Non-Discrete Questions	64
	Providing Non-Discrete Answer Option Sets	66
	Providing Uncomprehensive Answer Option Sets	68
	Leading Questions	70
	Social Desirability	71
	Thinking About the Questionnaire Structurally	72
	The Effect of Question and Answer Option Order	72
	Linking Questions for Analyses	74
	Validity and Reliability Concerns	76
	Conclusion	78
5	**Ensuring Survey Quality: Piloting, Checks and Cross-Cultural Comparability**	**81**
	Introduction	82
	The Complex Path to a Good-Quality Survey	82
	Pre-Fieldwork Quality Checks	87
	Cognitive Interviewing	87
	Piloting	89
	Interviewer Preparation and Briefings	90
	Translation	93
	Post-Fieldwork Quality Checks	94
	Data Cleaning	94
	Sensitivity Analyses	96
	Cross-Cultural Construct Validity	97

6 Conclusion 103

Glossary 109

References 111

Index 119

LIST OF TABLES AND BOXES

List of tables

2.1 Population statistics for Poland, sex and age
(Eurostat, 2016) and quota calculation   22

3.1 Influences on survey participation according to
Groves and Couper (1998, p. 30)   44
3.2 Calculating weights for a survey sample from Poland
with reference to population statistics (Eurostat, 2016)   47

4.1 Example of an 11-point scale with a labelled midpoint   61
4.2 Example questions from the World Values Survey (2012)   75

List of boxes

2.1 Case Study: Predicting the US Presidential Elections 1936: Why
Sampling Is So Important   11
2.2 Ask an Expert: Rachel Ormston | Sampling Narrowly
Defined Groups of Respondents   14
2.3 Ask an Expert: John Curtice | Surveys, Polls and Exit Polls   27

3.1 Case Study: Random Route Procedure in Wave 5 of the
German World Values Survey Sampling   35
3.2 Ask an Expert: John Curtice | Mistakes in UK General Election Polling 49

4.1 Ask an Expert: Good Survey Questions and Common Pitfalls   63
4.2 Case Study: Asking About the Independence of Scotland   73

5.1 Case Study: The Processes Required to Create the Scottish
     Social Attitudes Survey   83
5.2 Case Study: Young People's Political Engagement   86
5.3 Ask an Expert: Susan Reid and Paul Bradshaw: On Cognitive
     Interviews and Pilots   91

ABOUT THE AUTHOR

Jan Eichhorn is a Senior Lecturer in social policy at the University of Edinburgh and Research Director of the Berlin-based think tank d|part. His research, on political engagement and challenges to orthodox economic assumptions, is based on large-scale survey research. With a variety of partners, he has designed within-country and internationally comparative surveys on a range of topics using many different approaches to data collection. Dr Eichhorn's research has been featured extensively in print and broadcasting media, and he has provided briefings and consultancy for several governments, civil society organisations and political foundations. He has also taught survey methods to undergraduate and postgraduate students, civil servants and third sector workers.

1

INTRODUCTION

Chapter Overview

Why do we need to worry about sampling and survey design?.................. 2

The approach of this volume.. 3

Overview of the content of this volume.. 5

Further Reading 8

Why do we need to worry about sampling and survey design?

According to a report by a committee of the Scottish Parliament issued in 2015, something extraordinary must have taken place in the aftermath of the 2014 Scottish independence referendum. Youth participation must have increased dramatically following the lowering of the voting age to 16 years. The headline of the news item covering the findings of a survey conducted for this age group by the Scottish Parliament read,

> An overwhelming majority of 16- and 17-year-olds who were eligible to vote in last year's historic referendum did so, with four out of five saying they want a vote in all future elections, a Scottish Parliament committee survey has found. More than 1200 of the eligible first-time voters responded to the online survey run by the Devolution (Further Powers) Committee. (Scottish Parliament, 2015b)

Indeed, these insights matched the results from other research that suggested that those young people who had recently been enfranchised largely made use of the opportunity to vote (especially compared to 18- to 24-year-olds, who participated much less; Electoral Commission, 2014, p. 64) and were also showing greater political engagement than their peers elsewhere in the UK (Eichhorn, 2018). At first sight, everything seems to be fine. A survey of 1200 respondents would usually be seen as a robust sample, and the results seemed to agree with findings from other research too. However, once we look at the details a little more, we might become more sceptical.

One of the most remarkable findings of the survey, the increase in young people joining political parties, made national news headlines. *The Guardian*, for example, titled 'Quarter of youngest Scottish voters have joined a party since referendum' (Brooks, 2015). The author cited the findings from the Scottish Parliament survey, which said that 26% of 16- and 17-year-olds had joined a political party since the independence referendum (Scottish Parliament, 2015a). Let us break those figures down for a moment. According to the Scottish Government's records, there were roughly 122,000 16- and 17-year-olds living in Scotland in 2014 (Scottish Government, 2015). There was indeed a substantial increase in party membership immediately following the referendum, in particular for those parties that supported independence. The Scottish National Party's (SNP's) membership reached 93,000 in January 2015 (Keen & Apostolova, 2017) – up from about 26,000 on the eve of the referendum – while the Green Party's membership increased from just under 2000 to around 9000 (Kennouche, 2015) during the same period. If we take those figures together[i], this would correspond roughly to a total increase of party members of 74,000. If indeed

---

[i]Other Scottish political parties did not reveal membership figure changes in the same period and in informal interviews suggested that, while they also welcomed some new members, increases were in the hundreds rather than thousands.

26% of 16- and 17-year-olds had joined a party in this period, that would be equivalent to roughly 31,700. This would suggest that of all the people who joined political parties in the wake of the Scottish independence referendum, somewhere around 42% to 43% were 16- or 17-year olds. That already seems suspiciously high. If we then look at research into the actual structure of party membership, further doubts are cast, considering that the mean age of SNP members was approximately 49 years ahead of the 2015 general election (Bale & Webb, 2015, p. 6). This clearly could not be the case if so many 16- and 17-year-olds had joined. Something must have gone wrong here.

A look into the background notes on the Scottish Parliament survey provides the main insight: 'The survey was distributed to young people via schools, colleges and youth clubs. More than 1200 responses were received'. At the same time, the committee responsible ran a campaign to get young people in schools to engage with these issues and the Parliament. So the likely result of this was that the participants in this survey were composed in particular of school students who were either generally interested in politics or specifically engaged with related topics through classes in school or youth club activities. In other words, the young people surveyed were probably not very representative of their peers across Scotland more generally. The sample of 1200 respondents was skewed towards those who were more interested or engaged in politics already. This is often referred to as a self-selection bias in sampling. The news issued about the survey and the articles in the media reporting it, however, did not comment on this problem. So, quickly the image arose that young people in Scotland were even more politically engaged than they had indeed been found to be. The extent was massively exaggerated.

The approach of this volume

To avoid such mistakes, it is crucial to design a good sampling strategy when conducting any form of social survey. While we are, understandably, mostly interested in the results following the analysis of quantitative data, making sure that the data we get to work with are actually adequate is fundamental if we want to ensure that we can make statements that reflect what is really going on. It is also crucial for us to understand this in order to evaluate the quality of other research when assessing how meaningful the insights from it really are. This volume is about the part of the research that takes place before we get any quantitative data that we can actually analyse. It is about the stage at which we design the research, in terms of how we sample relevant groups of people for our studies, how we construct questionnaires that enable us to ask people about the issues we are interested in and how we check the quality of our data before we begin to analyse them properly. None of these tasks are easy, and they require a significant amount of critical thinking. The design of a survey project can look very different depending on the scope and aims – from quick polls that need to be administered within days to complex surveys that may take over a year from start to finish. Speed is one issue that can affect the process. Additionally, budget constraints often play an important role. In most situations, we have a good idea of how we could maximise the quality of a survey from a theoretical point of view; however, the costs are often prohibitive. Therefore, we need to find ways to achieve good quality outcomes even when our resources are constrained.

This volume aims to provide an overview of the most important issues that we need to take into account when designing surveys and developing sampling strategies to achieve high-quality outcomes. It will familiarise the reader with the variety of different approaches that can be chosen and discusses advantages and disadvantages of them. Crucially, this volume is meant to be applicable to the real world and the practice of carrying out this work. As the example above illustrates, the approaches we choose for collecting survey data heavily impact to what extent the analyses we conduct later on can be seen as valid and meaningful. Therefore, we will not only look at theoretical considerations – for example, with regard to what sampling strategies may be likely to provide the most representative types of samples – but also ask how we can achieve the best quality research while keeping in mind the real-life constraints that are often placed on researchers in practice (e.g. time, budget and topical limitations). Crucially, this volume is introductory and does not assume prior knowledge on the part of its readers. It should be accessible to any person who wants to learn about the topics covered. The goal is to provide the reader with access into the debates about questionnaire design and sampling and raise awareness about important aspects researchers should consider when they look at the data produced by others. Each chapter in this volume could have a whole volume dedicated to it in its own right, however. In order to capture a wide range of issues that we should consider when designing surveys, the depth we can go into regarding each individual topic has to be limited. On many occasions, other volumes in this series provide those deeper insights, and where that applies, we suggest such links. Also, in the chapters, we cite work by methodological experts exploring the respective topics and would encourage readers to follow up those references if they are interested in finding out more details about the specific themes discussed. The principal goal of the volume, however, is that readers will be able to engage with discussions about sampling and questionnaire design and the implications that stem from both for survey data in a meaningful way, both in terms of the academic background and in particular also the practical side of it.

To achieve this, we will not only discuss sampling and survey design in an abstract manner but apply those discussions to real-life examples. The chapters in this volume contain case studies about specific survey projects and discuss shortcomings in surveys and polls where results may have been distorted on the one hand and highlight examples of good practice and insights from high-quality approaches on the other hand.

The surveys referred to range from smaller scale regional to large-scale comparative international surveys. Some focus on a very specific subgroup of the population, while others try their best to represent the general population of countries. Using those case studies will help to establish the relevance of the theoretical concerns raised in this volume and enable all readers to consider possible applications following some reflections. As survey research is carried out by people, many of whom specialise in this field and spend many years continuously improving their knowledge and abilities to undertake the best possible work, we will further complement our insights by hearing from some of those professionals. To gain an insight into the practical aspects of survey design and sampling work, this volume will present insights from interviews with professionals specifically carried out for this volume. To show the breadth of applications, we interview experts from academia, professional research institutes and the private sector, reflecting the wide range of contexts in which survey work is carried out. The interviewees for the volume are as follows:

- Paul Bradshaw, who is the Director of ScotCen Social Research and has previously been Group Head of Longitudinal Surveys for NatCen Social Research
- Prof. Sir John Curtice, who is an expert psephologist, Professor of Politics at the University of Strathclyde and the President of the British Polling Council
- Rachel Ormston, who is Associate Director at Ipsos MORI Scotland and who has previously worked for multiple survey and polling organisations
- Susan Reid, who is Research Director (Social Attitudes) at ScotCen Social Research and Project Manager of the Scottish Social Attitudes Survey
- Prof. Dr Christian Welzel, who is Professor for Political Culture Research at Leuphana Universität Lüneburg and Vice President of the World Values Survey Association

The interviews with the interviewees helped to prioritise key issues for discussion in this volume, but most importantly, their contributions, as insights from practitioners, are integrated directly into the context of this volume, most notably in the form of 'Ask an Expert' sections, where we present their accounts directly in relation to the topics of the respective sections.[ii]

Overview of the content of this volume

In this volume, we engage extensively with questions related to the development stage and the data collection forming the foundation of social surveys. So, it is crucial

---

[ii]The volume presents direct quotes from the interviews. They are taken verbatim, but have been corrected for grammar mistakes and filler words and phrases have been removed. The edited quotes used here have been shown to interviewees for confirmation of consent.

to understand what the purpose of sampling is and how we go about it. In Chapter 2, we address this question.

We discuss in detail how we should approach the sampling process, so that the people we interview as respondents actually allow us to make statements about the population they are drawn from. To do this, we introduce the logic that underpins sampling for social surveys. Our main aim is to establish a sample that is representative, and we will discuss what that precisely means for a research design. The understanding will be aided by engaging with the concept of a **sampling frame**, which provides the set of possible respondents who could be interviewed. There are many different ways how this can be undertaken, and readers will learn about the difference between probability and non-probability sampling methods, their advantages and disadvantages and how they relate to questions of representativeness. In both instances, we consider multiple subtypes (e.g. stratified, multistage and cluster sampling in the case of the former, and quota, convenience and snowball sampling in the case of the latter). A classic case study into the 1936 US presidential elections illustrates the importance of understanding sampling methods well. Two 'Ask an Expert' sections provide some deep dives, first, to learn about the specific problems of sampling very small and specific population groups (e.g. young people or certain ethnic minorities), and second, to explore the distinction between surveys and polls and the special characteristics of exit polls, which we encounter on election nights.

After engaging with the theories behind sampling methods, Chapter 3 looks at the implementation in practice. Even within a particular sampling methodology, there are many ways of delivering the data collection. Interviews with respondents can be carried out face to face, over the telephone, online using a variety of platforms or via physical mail, for example. The mode of conducting survey interviews is important, as it can affect how people respond. People may give slightly different answers to the same question if asked in person or when entering their reply on a computer screen. In this chapter, we discuss the advantages and disadvantages of different ways of implementing sampling methods for the data collection, considering both quality-oriented concerns and very practical ones, such as budget and time constraints, and importantly, what to do when reaching particular groups of respondents is very difficult and why it can be helpful to combine multiple modes of data collection. We also briefly discuss why small deviations from perfect distributions in samples may not be a problem if we can make meaningful adjustments through a process called weighting. In a case study from the World Values Survey's implementation in Germany, we look at how physically interviewers in face-to-face administration modes actually go about recruiting respondents, helping us to understand why such methods are so research intensive. We apply the knowledge gained in an 'Ask an Expert' section, where the misevaluations in polling ahead of recent UK general elections are discussed.

Getting the right sample of people to take part in the survey is crucial, but of course it is only the, albeit very important, starting point of a survey project. We also need to be able to ask people good questions in order to make sure that the information we receive actually enables us to conduct the analyses we are interested in. Chapter 4 therefore focuses on questionnaire design. We discuss different strategies for the development of questions and distinguish different types. Subsequently, we discuss how such a process can be undertaken and consider how we might think of questions not just individually but as grouped sets depending on the concepts we plan to study. Afterwards, we look at what makes a good question and what common pitfalls exist that might make a question less useful or even invalid. The chapter introduces readers to different types of questions and makes them aware of how these different varieties of question constructs correspond better or worse to certain types of analyses, which is why it is so important to think carefully not only about the content but also about the structure of the questions. Our expert interviewees reflect about what they deem to be the most important characteristics of good questions and what they think are the most common mistakes in questionnaire design that should be avoided. The case study in this chapter places us in the heated debate about Scottish independence in the lead-up to the referendum in 2014 and discusses how question order effects have been used to produce a particular polling result for a political campaign.

Even with the best planning and the application of existing theoretical and practical knowledge, it is often very difficult to know how exactly a survey question will work when it is used in the field. This particularly applies to new questions that have not been used before or those that are applied to contexts in which they have not been used previously. Chapter 5 discusses a very important stage in the development of surveys: piloting. Especially when we consider how costly and time-consuming surveys can be, it is important that we ensure that they will work in the way they are intended to before rolling them out fully. In this chapter, we therefore describe different options for testing how well a survey works through a number of piloting strategies, including the utilisation of internal checks, cognitive interviewing and small-scale tests. Furthermore, we discuss how interviewers can be prepared to minimise problems in implementing a survey and what limitations there may be to the feasibility of extensive piloting. Finally, we also look at what initial checks researchers can do during the pilots and when they receive the data sets after fieldwork to assess the quality – before any of the actual analyses can begin. The chapter also pays attention to the specific challenges of designing and administrating surveys across different countries, by discussing the issue of translation and cross-cultural comparability. The insights from our expert interviewees are particularly helpful in illustrating how those techniques are implemented in practice, so we not only discuss the useful feature of cognitive interviewing with them but also ask about the main insights that we can gain from pilots and how pilots differ between offline and online administered surveys. Using the Scottish Social Attitudes Survey in a case study, we explore the many steps that have to be undertaken to actually bring together sampling and questionnaire design with all the other features that are necessary in order to be able to conduct a high-quality survey.

In the final chapter, we bring together the main lessons learnt from the theoretical and practical discussions throughout this volume. In addition, we outline how the issues learnt here relate to other aspects of quantitative methods. That way, readers will be able to identify how sampling and survey design explicitly affect the analysis of the data ultimately generated. Furthermore, we will look at how this knowledge is highly useful in assessing empirical claims made in quantitative studies in academia and outside of it.

Further Reading

Silver, N. (2015). *The signal and the noise: Why so many predictions fail – but some don't*. Penguin Press.

This is a great and accessible text for anyone who wants to know not only how evidence based on surveys of different kinds can be used meaningfully but also what sort of pitfalls often arise in real-world applications.

2

SAMPLING APPROACHES: HOW TO ACHIEVE REPRESENTATIVENESS

Chapter Overview

Introduction 10

Populations and sampling frames 12

Probability and non-probability sampling 15

Approaches to probability sampling 16

Approaches to non-probability sampling 20

Further Reading 29

Introduction

The whole goal of survey samples is to enable us, using quantitative methods, to analyse social scientific questions in relation to large groups of people – thousands or even millions of them. If we wanted to gather data about each person, for example, who is registered to vote in the national elections of a particular country, say Japan, we would have to talk to an unfeasibly large amount of people, in this case more than 100 million (International Foundation for Electoral Systems, 2018). The amount of time and resources this would require is prohibitive. So, what is the next best thing that we can do instead? We can try to identify a smaller group of all the registered voters that is in its characteristics close to identical to the characteristics of the group we are interested in. **Sampling** then allows us to find a compromise between the degree of accuracy we aim to achieve with regard to a question we have about a large group of people and the resources that we have available to address the question (Stephan & McCarthy, 1958, p. 12). Sampling is complicated and requires a lot of thinking and the making of difficult decisions between a great variety of possible approaches. When considering all aspects that we have to take into account when engaging with sampling, it can seem quite daunting, in particular when our theoretical ideal cannot be achieved in practice, because of limitations that may hinder us to access those respondents we would ideally like to reach. However, the effort is worth it, because sampling allows us to do something very powerful. Using robust approaches to selecting a sample and developing meaningful surveys, we are able to make statements about characteristics of groups much larger than our sample.

If we are confident that the composition of our smaller *sample* is then approximately similar to that of the larger group (which is referred to as the *population*), we can conduct analyses with only our sample, but we are able to make statements about the population as a whole, with a degree of certainty. There will be some margin of error around the estimates from our sample, but within that margin we can be confident that we can talk about our population as a whole. For this to work, it is crucial though that our sample indeed reflects the characteristics of the population. If it does not, any analyses of the sample only reveal to us what the sample looks like, but we could not generalise to our population. So, it is important to 'get the sampling right', if we want to make statements about our target population. That is why we need to pay close attention and understand how the decisions about our approach to sampling affect the results thereof. In this chapter, we will introduce the most important approaches to sampling. However, before we begin to discuss the samples, we first need to be clear about what constitutes our target population.

Box 2.1: Case Study

Predicting the US presidential elections 1936: Why sampling is so important

(Based on Squire, 1988)

In 1936, the *Literary Digest*, a popular magazine in the USA, conducted a poll to predict who would win the presidential elections that year – the incumbent, Franklin Roosevelt, or the main challenger, Alf Landon. The magazine had correctly predicted who would win the elections on all previous occasions since 1920 and prepared a large-scale effort to do this again. They sent out more than 10 million ballots and received more than 2.2 million returns that they counted. The selection of their sample was based on entries in car registration lists and phone books. Such a large sample would initially instil a fair deal of confidence in anyone reading about it without any knowledge of sampling methods. However, the result was not marginally but categorically different from the real results. While the magazine's poll suggested Landon would win with 55% of the vote, followed by Roosevelt with 41%, the actual results on election day saw Roosevelt re-elected with a massive 61%, while Landon received only 37% of all votes cast. At the same time, based on a competitor survey conducted by Gallup, Roper and Crossley predicted Roosevelt's win correctly (albeit not with the perfect percentages). That survey, however, had only a few thousand respondents. How could it be that such a small survey could outperform such a large sample?

The crucial answer lies both within the design and the actual undertaking of the sampling process. Good surveys measuring national public opinion can commonly achieve meaningful results with rather small samples of 1000 to 2000 respondents, while surveys with much larger samples, but with much poorer sampling designs, may produce results that are not representative of the real population at all. The 1936 *Literary Digest* survey suffered from two important shortcomings. First, the initial sample was biased and included a disproportionate amount of Landon voters. In the period following the Great Depression, one suggested explanation was that basing a sample on car and telephone registration lists biased the sample against people who were hard hit by the depression and may not have had access to either good. That effect may have been stronger than the effects in previous elections, before and in the early phase of the economic crisis, where fewer people had experienced strong negative impacts that may have distorted the sample. Second, Landon supporters were more likely to respond to the questionnaire, and therefore, when looking at the returned straw ballots, an overestimation of the support for Landon was registered.

Sample size is important when conducting survey research, but the quality of the composition of the sample is even more fundamental. Regardless of the sample size, a poorly constituted sample will lead to potentially heavily biased results. Understanding the different approaches to sampling and their advantages and shortcomings is therefore crucial to the design of useful surveys.

Populations and sampling frames

At first, it may sound strange and nearly trivial to say that identifying a target population carefully is of paramount importance. Is it not usually rather obvious what we want to study? Actually, it can be quite tricky. Let us imagine that your overarching goal is to investigate public attitudes in a recent election within a country. Who comprises the set of people that should form part of your investigation? Is it everyone who lives in a particular country, because they are all affected by what is being decided by the politicians elected to office? Or should it only be about those people who are actually eligible to vote, as you may want to focus on the behaviour of the electorate specifically, rather than the perceptions of people in the country more broadly? If that is the case, do you need to include people who have the right to vote but live outside the country of interest and take part through external voting (e.g. in embassies or by international mail)? Could it be that you are actually only interested in the behaviour of people who took part in the election and, therefore, you should exclude people who are eligible to vote but who decided not to participate?

Something that might sound like a straightforward theme, such as an election, can be looked at in many ways, even when focussed on public attitudes. It reminds us that, as for any good research, we must start with a well-defined understanding of our goals and, ideally, a clear research question. Once we have developed it, we can actively make a decision about who needs to be included in our conception of the population for our study and who should be excluded. The population should include everyone who matches the criteria required by our question. In order to do this well, Sudman (1976) suggests a two-step process. First, we should identify the units of your analysis, and second, define the characteristics that need to apply to them.

With regard to the former, survey research is often interested in individual persons. But that does not always have to be the case. Our unit of analysis could not only be families or households, for example, but also organisations (e.g. firms), organised groups of individuals (e.g. initiatives or community groups) or transactions (e.g. exchanges in a particular market place). After identifying that clearly, common characteristics that may further describe the structure of our sample may include geography (e.g. if we are only interested in the electoral results in a particular part of the country), the age of persons (e.g. if we are investigating first-time voters only), other demographic variables (e.g. gender, marital status or education) or a range of measures that could reflect personal background or household composition (Sudman, 1976, pp. 12–13).

When we try to emulate the population in our samples, it is important that we only formulate insights that speak to the population our sample is designed to represent.

This is very important and often a challenge. The broader and more complex the population is, the harder it can be to derive a sample that genuinely matches its composition. Conversely, the more specific and narrow the parameters defining our population are, the more difficult it can be to actually identify potential respondents for a survey. We see the former concern, for example, often prominently raised in experimental studies based on samples composed of students, who can easily be accessed by researchers at universities (Kam et al., 2007, p. 416). But, obviously, there are many reasons to question whether a finding based on a group of university students is reflective of a population that is conceived of as encompassing more groups than university students in a particular place (Sears, 1986) – though some research suggests that such student-based samples may often provide equivalent results to non-student samples (Druckman & Kam, 2011). These considerations also act as an important reminder that findings from respondents in one cultural context may not be translatable directly to people elsewhere in the world, so even if a sample is representative for the population of a country more broadly, we should be cautious to generalise with regard to people more widely (Henrich et al., 2010). This is an issue we will discuss more in Chapter 5.

The latter problem may be particularly prominent if we specify a population that is very small, but not linked together through a simple structure of comprehensively identifiable institutions or organisations (e.g. female vegans aged 45–60 years). These considerations highlight the difficulty in identifying a suitable so-called *sampling frame*. After defining a target population, we obviously want to create a useful sample thereof. In order to achieve this, we need to have access to all people who make up that population in the first place. That, however, is often easier said than done. While it may be fairly straightforward to identify a list of all participants in a programme at a given university, for example, it could be very hard to identify a list of contacts for all university students in a country overall. While official databases exist for certain populations in some countries, such as electoral registers, properties of other registers may be less well known. Even when we have a register, it may be incomplete and particularly under-represent certain groups of the population (see Volume 1 in this series for a discussion of non-coverage and other survey errors). Even if full sampling frames exist in certain registers, they may not always be accessible to everyone straightforwardly, highlighting several of the practical issues that may arise in finding a suitable sampling frame that corresponds to our target population (Moser & Kalton, 1971, p. 48).

Rachel Ormston, one of the expert interviewees for this volume, highlights the difficulties in sampling very specific groups, looking at surveys of young people on the one hand and minority ethnic groups on the other hand in the following discussion.

Rachel Ormston | Sampling narrowly defined groups of respondents

What approaches work best to reach very specific, smaller groups of the population, like young people or ethnic minority groups?

With young people you can obviously do school-based surveys if you're looking at school-age young people, but then you have to take into account the fact that that will miss out young people who are not in education. . . . Also I think you have to think about subject matter, because there's been some quite interesting studies around asking young people about crime. If you ask them in a survey at school you get different responses because they feel different about answering that in different contexts. With minority ethnic groups that's really difficult actually, and there are different things that have been done, but some raise ethical issues. The most robust way you could do it generally would be focused enumeration, where you randomly select addresses but you get the interviewer to ask about the addresses on either side. So, they're basically screening say five addresses for every one address they visit to try and identify whether there's anyone from a minority ethnic background who lives there. But obviously that then raises questions, because the interviewer is asking questions about your neighbour's ethnicity. So that has to be done quite carefully. There are studies that have done it based on surname, which raises similar kinds of ethical issues because you're making assumptions about people's ethnicity based on their surname, which is obviously quite a crude tool. You can use existing sample frames, such as online panels that will have data about people's ethnicity, to sample based on ethnicity, but obviously that's not going to be a completely pure probability sample, because they're opt-in panels basically but they are ethical given they've provided that information voluntarily. You can argue that that might actually be a more ethical way of doing it than screening a surname. If you want to do it well – it's really expensive and quite ethically difficult.

If you had unlimited resources, what would be your ideal approach?

I would do focused enumeration, because if you really want to get an estimate of a sort of population level, I think all of the other approaches will just leave out too many people. But I would do focused enumeration where you probably start by sampling say census output areas that are known to have a higher proportion. So you oversample areas where you know there are more people from minority ethnic backgrounds living there. But do it so that you still have some areas where actually there aren't that many people and you have to try a bit harder to find them.

Are there any approaches for sampling small, specific groups that you'd say are absolute no-gos?

While you might use it in qualitative sampling . . ., but if you're snowballing from people, asking respondents, 'Do you know anyone else who fits in this group', you're going to get just a very homogeneous sample which probably isn't particularly representative. That said, a lot of charities will use that kind of approach. They will email out to their database of people who support them or are engaged with them in some way. Say a young persons' advocacy group might do that to all the people on their database and then that would be their survey of kind of 16- to 24-year-olds, which I don't think I would say that it was a no go, I think I would say it's fine, but you just have to present it as this is a survey of 16- to 24-year-olds who are in contact with your organisation, which means that they are likely to be slightly more kind of activist, slightly more concerned about the sorts of issues that they're being asked about already. So that's not necessarily representative of all 16- to 24-year-olds, some of whom might be much less bothered about some of the issues that they're being asked about . . . [These organisations often] say, 'we asked 16- to 24-year-olds'. But I would say, 'no you didn't, you asked 16- to 24-year-olds on your contact database who are likely to be quite different from the population of all 16- to 24-year-olds in ways that might be quite significant.'

Fundamentally, defining our target population carefully and identifying an appropriately corresponding sampling frame are crucial initial steps for any survey research project before we can begin the process of sampling. While it may appear straightforward initially, these decisions are not trivial and require careful attention in order to assure that the analyses conducted actually allow the researcher to make statements about the groups of people, institutions, organisations or transactions that they are interested in.

Probability and non-probability sampling

We distinguish two main approaches to sampling, probability and non-probability. In survey research, probability approaches are classically seen as the most desirable way of achieving a representative sample. Fundamentally, they provide a sampling mechanism in which each member of the population has a specific and known probability of being selected into the sample. This allows us to design sampling processes in which the likelihoods of any particular sample composition can be estimated, and therefore, we can calculate the probability of our sample results being representative of the population as a whole, although in the sampling process, we do not need to take account of any characteristics of the particular individuals being selected. The most direct applications of this approach are random sampling techniques, which will be discussed below and which will enable us to understand this rather abstract principle more clearly. However, such techniques cannot always be applied directly, mainly because of the lack of suitable sampling frames, which is why we will also consider other techniques (specifically multistage, stratified and cluster-sampling approaches). There are also sampling techniques that do not follow a probability-based design and instead rely on other techniques. In particular, in certain areas of polling work, non-probability sampling is often applied, usually through some form of quota sampling. In that approach, a researcher aims to actively mimic the population by measuring a range of characteristics within the sample to try to maximise their similarity to the population overall. Below, we will discuss this in more detail and review their advantages and disadvantages.

Approaches to probability sampling

Simple random sampling

Kalton (1983) usefully notes,

> Simple random sampling (SRS) provides a natural starting point for a discussion of probability sampling methods, not because it is widely used – it is not – but because it is the simplest method and it underlies many of the more complex methods. (p. 9)

In other words, simple random sampling represents the conceptually most straightforward approach to using probability-based approaches in obtaining a representative sample. However, as we will see, it is very difficult to execute in practice in many instances, which is why we use its logic as the foundation for other techniques.

Simple random sampling is rather straightforward indeed. Starting with our target population, we define a certain sample size, and we will select that number of respondents for our survey from the population through a fully randomised procedure. In practice, this means crucially that each individual within the target population has the exact same chance of being selected into our sample. If that can be achieved, we are in a very strong position. By virtue of randomisation, we could expect that our sample, as it increases in size, would become more and more similar in its distribution to that of the population. Think about a simple example of a coin toss, where you have two possible outcomes, heads or tails. Each has an equal chance of 50% to occur. If you toss the coin only a few times, it is fairly likely that the distribution may be skewed towards more heads or more tails, but if you flip the coin very often (and nobody tampered with the coin), say 1000 times, you would expect that overall the imbalances would roughly even out and you would get a result overall where just less than or more than 500 tosses would result in heads and tails, respectively. So without actively manipulating your sample of coin tosses (your target population here being the infinitely many coin tosses that could be undertaken) in any other way, merely by randomisation, you would be able to achieve a sample that would likely be close to the distribution of the population overall (50/50). While it might not be perfect, because of the known probabilities, we are actually able to calculate the likelihood of our sample results being equivalent to the results in the population. The logic applied is that of inference – however, we are not dealing with that in this volume. It is covered extensively in its own volume (Volume 3) as part of this series.

So why then do we not simply use simple random sampling methods all the time, if they seem so ideal and straightforward? They can even be easy to implement: imagine you wanted to generate a random sample of 100 of all the members of the UK Parliament (House of Commons and House of Lords combined). All you would have to do was to get the list of all their names (which is publicly available) and create a random selection mechanism. You could do this (quite tediously) in a manual fashion, for example, by writing each name on an equally sized sheet of paper, mixing those up well in a closed box and picking out 100 of the sheets of paper. More commonly nowadays, you would probably use a computer program that would replicate this process of random selection from your list. But the only reason you can do this in the first place is that the list of all members of that target population is actually easily available. For it to work, you need to have a complete sampling frame. That, however, is unfortunately often not the case when we want to undertake social surveys. If you wanted to conduct an attitudes survey of all adults in a particular country, you would already begin to struggle. Even if there is a register of every person in a country with their home address and that register was kept up to date consistently, it is very unlikely that the state institution holding that register would permit anyone to access it for their own research needs. But without a sampling frame, you cannot apply any randomised selection mechanism in practice, because you simply do not even know the names of all the individuals who form your target population (so you could not make a set of sheets to put in a, in this case very large, box to randomly pick from).

This has an important practical implication for implementing random sampling-based techniques. We need to make sure that the respondents who were randomly selected are indeed the ones that are asked the questions of a particular survey. Imagine, for example, that we had a sampling frame with the telephone numbers of the target population. We would call a randomly selected number of people on the list. But not all respondents will be available the moment we first dial the number. Should we then simply move on and call someone else? No, we should not! If we did that, we may create biases in our sample (maybe some groups of the population are more available to be reached over the phone, for example). Instead, we should make repeated attempts to contact the originally selected person to make sure that our random sampling approach properly works (we will discuss how to properly conduct the data collection in more detail in Chapter 3). So even when we have a clear sampling frame, random sampling requires a lot of effort in practice.

Often, however, we do not have a straightforward sampling frame at all. In those instances, we need to apply other methods to still obtain probability-based samples. We will discuss some of the most prominent approaches below.

Cluster and multistage sampling

Often, we may not have the information for all potential respondents in our target population for a survey; however, we may have information about aggregations of these respondents that we can utilise. In multistage and cluster sampling approaches, we may not be able to create a sample from the full list of our population, but we may be able to generate a random sample of groups of our respondents as a first step. A very common example where this approach is applied to is in relation to studying school students. Even if there are registers for all school students, researchers would be unlikely to be granted access to those because of data protection. However, a list of all schools in a particular region or country of interest may very well exist, thus providing a researcher with a route to comprehensively identifying all possible locations where the ultimate members of the target population (school students in this case) could be found.

What we effectively do is divide our total target population in several defined sub-clusters. These clusters have to fulfil certain conditions (Arnab, 2017, p. 409). They should be comprehensive (i.e. they should cover all respondents from the population – in our case, all students in a particular country) and they have to be mutually exclusive (i.e. each respondent can only be part of one cluster at a given time – in our case, this means each student should only be enrolled in one school). If these conditions are fulfilled, we can then draw a sample, following the approach outlined above from random sampling techniques, of all schools in the country in the first instance. Subsequently, we could then include all the students within each cluster in our sample. In a sense, we have moved the randomisation one aggregation level up. We obviously need to sample a sufficiently large number of schools, as schools themselves will differ with regard to the composition of their students, but the general logic of randomisation still applies and we can continue to work within a probabilistic framework that permits us to use inferential logics – albeit having to take into account some caveats regarding the calculation of estimates following the different sampling process (Kalton, 1983, pp. 29–38) and being able to explicitly investigate not just individual-level effects but also differences between clusters (which can, in the first instance, form part of the research question).

However, it may not always be feasible to include all respondents within all clusters within our sample in the final survey. If we have selected a large number of schools, for example, the total number of students may be very large and beyond the scope required for the analyses in a project. It may also be organisationally difficult to organise written consent from parents as part of the process for everyone, while it could be more feasible to arrange for these things to be done at the class level rather than overall school level. So we may have more stages of identifying sub-clusters or indeed only select a random sample of respondents within the final clusters we have identified. While the terms are sometimes mixed in the literature and generally referred to as cluster sampling (Kalton, 1983, p. 29), to be precise, we can differentiate between cluster sampling in the narrow sense and multistage sampling. While *cluster sampling* specifically refers to the sampling of clusters and then the selection of all individuals within those clusters selected, *multistage sampling* permits that within selected clusters a further sampling process takes place that will result in only subsamples of the initial clusters being selected (Arnab, 2017, p. 423). Crucially, at each stage the sampling processes should follow randomised processes as closely as possible, so that the probabilistic approach can be retained and inferential logic applied.

Subsequent analyses that utilise data from multistage sampling designs need to take the complex structure of the data into account carefully. As stated above, there may be systematic differences between our clusters in the first place. For example, some schools may have more students from higher income families than others. Therefore, the individual respondents (here the students) are not fully independent of each other. This applies at all levels of the sampling. The performance of students in a standardised test may, for example, be affected by the quality of the teaching they have been exposed to, so there may be group effects when comparing one class to another, because they were taught by different teachers. So in analyses of survey data originating from cluster or multistage sampling, we need to take into account these complexities and potential clustering patterns that are a consequence of subsamples of respondents being somehow connected to each other through shared experiences. One common approach that utilises the complexity of information originating from such samples is multilevel modelling – a technique discussed extensively in Volume 8 of this series.

Stratified sampling

Cluster sampling approaches can be particularly suitable when we try to recruit respondents within a clearly defined group of the population. But even when we are trying to engage with much broader populations (e.g. all adults in a country), it can be very helpful to break down our analysis into certain subgroups. Quite often, we may actually have some knowledge about the composition of our target population and different constituent parts. It may be easier, for example, to undertake sampling processes within each of a set of administrative regions or distinguish the ethnicities that people may have, in particular if our research is about comparisons between different groups of a certain characteristic (e.g. geography or ethnicity). Stratification allows the researcher to define the subpopulations or *strata* to then draw a sample within each of these strata separately. This can be very important when we expect, for example, that a whole-population approach to random sampling may actually result in biased samples, because response rates may not be uniform across different groups. People in a particular region or who have a particular ethnicity may be more or less likely to take part in surveys, for example. Stratification can ensure that the sample size for each relevant group is therefore completely reached. The most crucial point then is that sample sizes for subgroups are not left to randomisation but are controlled by the researcher (Kalton, 1983, p. 20).

Researchers can choose to match the sample sizes proportionately to the distribution of the different strata in the target population (referred to as proportionate stratification). However, sometimes, researchers may intentionally deviate from this to increase the sample size of a particular strata beyond the number you would expect from whole-population random sampling (disproportionate stratification). This is commonly done when the research question asks for the engagement with a group that is relatively small in terms of the whole population and if left to random sampling, we would only get a very small sample of them that we could analyse – potentially too small for the investigations we would want to undertake. Therefore, we may oversample such groups intentionally, and stratification-based approaches allow us to do this. Apart from this strength, the ability to estimate characteristics of distinct subpopulations, Arnab (2017, p. 214) identifies several other advantages of stratification: administrative convenience, in making the sampling more manageable through applications to subgroups; improvements in the representativeness of the sample, in particular if certain subgroups may otherwise be underrepresented, due to, for example, differences in response rates, as mentioned above; efficiency in the estimation of group characteristics under scrutiny; and improved data quality if, for example, different investigators can carry out the data collection for different subgroups based on the language they speak.

Approaches to non-probability sampling

While probability sampling methods have many advantages, as outlined above, in particular as they are underpinned by a theoretical approach that focusses on minimising biases in the selection of cases and thus permitting the estimation of our confidence in the strength of the relationship between the sample and our target population, these approaches also have a major shortcoming: when focussed on large populations, such as people residing in a city or even a whole country, they are very expensive to administer. We will look at the different modes of practically collecting the data in Chapter 3, but to implement randomisation procedures for thousands or even millions of people is very labour-intensive, onerous and can take a long time. However, researchers do not always have such extensive resources or the time to undertake the work. For example, if a newspaper wants to conduct a quick opinion poll to find out whether people like a newly elected party leader, they need a response within a few days to report on it in a timely fashion and cannot wait for, what could sometimes be several months to undertake the data collection for certain types of extensive probability sample–based surveys (as we will see in Chapter 5). Non-probability sampling, in particular quota sampling, is therefore a rather commonly used method, especially in market and certain types of polling research. We will discuss below how it works and engage with the particular problems that need to be considered when employing it, before also briefly considering further alternative forms of non-probability sampling.

Quota sampling

The basic idea behind quota sampling is explained fairly straightforwardly. It takes our general starting point that we want to generate a sample that mirrors our target population well literally and tries to proactively recruit respondents that match the characteristics of the population. The process begins by deciding what characteristics most crucially describe the population we want to study. For example, in any sample of the adult population of the country, we would all quickly agree that it would have to contain both men and women, and it would also have to cover the different age groups of the population. Commonly (unless, again, one wanted to actively oversample a group), a researcher would try to design the structure of the sample according to the proportions in the population (Kalton, 1983, p. 92). So say, for example, that 51% of the adult population were female and 49% were male, the researcher would decide to use the same proportions as the targets for their sample as well. So if the sample had 1000 respondents, the aim would be to recruit 510 women and 490 men into the sample. Similarly, quotas for age would be set as well and the complexity of the quotas could be increased further through linkage. If they do not get linked, there would be a danger that, for example, we could get the right number of men and women, respectively, and the right number for each age group, but nearly all the women were in younger age groups and nearly all the men in older age groups – which would make the sample very unrepresentative in a substantive sense, of course.

Consider the following example. In Table 2.1 you see the distribution for sex and age in Poland based on 2016 Eurostat data. If we simply recruited respondents and made sure that each quota was filled, we would not be able to control in any way whether the sex distribution applied actually within each age group. So, assuming that we did not apply any other measures to address this, we could theoretically end up with a heavily distorted set of recruited respondents that would nevertheless be in compliance with our initial quotas. Therefore, what we would need to do is calculate so-called cross-quotas, which in this case, for simplicity, apply the sex ratio to each of the age groups, thus giving us much more fine-grained quotas. This would enable us to create a sample that in relation to sex and age would look very similar to the population overall. A researcher recruiting respondents would begin any interview by screening respondents for their age and sex, and if they fell into a category where the target was reached already, the survey would not be administered for them any more.

Table 2.1 Population statistics for Poland, sex and age (Eurostat, 2016) and quota calculation

Demographic group	National population distribution	Raw quota for the sample (total = 1000)	Sex	Age	One possible scenario based on raw quotas only	Estimation of cross-quotas
Male	48%	480	Male	18–25	20	62
Female	52%	520		26–34	30	86
Age (years)						
18–25	13%	130		35–44	90	82
26–34	18%	180		45–54	100	77
35–44	17%	170		55–64	110	86
45–54	16%	160		65+	130	86
55–64	18%	180	Female	18–25	110	68
65+	18%	180		26–34	150	94
				35–44	80	88
				45–54	60	83
				55–64	70	94
				65+	50	94

In many ways, quota sampling is quite intuitive and is very straightforward in its administration. So why do we not simply use it all the time, when probability sampling is so resource-intensive? It is because we do not benefit from the main advantage of probability sampling. Probabilistic methods, as shown earlier, allow us to obtain a sample that is representative of the population, because of the randomisation techniques we apply. We do not design the sample to have particular characteristics, the sample develops those characteristics (which are similar to the population), because of the technique (unless there are biases, e.g., in response rates, which we return to in Chapter 3, but which could affect any sampling method). When we apply quota samples, we only mirror the population distribution according to the quota characteristics, which we have decided upon as being relevant. However, we do not know whether there are other relevant distributional factors, which we are not taking into account.

Consider our example above again. While the distribution of sex and age would be identical to that in the population overall, we do not know what sorts of men and women, respectively, at each age group we recruited into our sample. Societies are stratified according to income and socio-economic class, for example. It is possible that our sample may include respondents who are disproportionately wealthy across all categories, or it may be unevenly biased, including men who are on average less wealthy than men in the population and women who tend to be more wealthy than the average we would expect to see. In order to ensure that this was not the case, we would also have to add quotas for these characteristics and supplement them potentially with further cross-quotas. But where should we stop? Do we need quotas for education, housing tenure, marital status, the number of children and religious affiliation, to name just a few? Crucially, many of these characteristics are related to each other. If we know a person's age, sex, income, education and housing tenure, we actually have a lot of information that permits us to predict a wide range of other socio-economic variables fairly well. But it fundamentally depends on the variables of interest we want to analyse in our survey. If there are particular characteristics that are strongly linked to those variables that are at the core of our analyses, it would be particularly pertinent to ensure that those are reflected well in the sampling distribution. However, this can be very difficult. If, for example, you are interested in analysing vote choice in an election, there are many factors that influence the outcome and how they influence it may change over time, as well.

It means quota sampling always has to rely on a degree of 'subjective evaluation' (Kalton, 1983, p. 92) that is not required in probability sampling methods. The problem could be observed, for example, in the 2015 UK general election, where most polls predicted a hung parliament requiring coalition government to continue, while in fact the Conservative Party ended up winning a majority of seats in the House of Commons. While several issues contributed to the incorrect estimations by most polling companies before the election, an inquiry by the British Polling Council and the Market Research Society found that the dominant reason leading to the failure was 'unrepresentative samples' (Sturgis et al., 2016, p. 4). They found that, crucially, support for the Labour Party was systematically overestimated, because groups that supported the Labour Party were over-represented following the application of the quotas used. It demonstrates the difficulty in applying quota sampling designs to topics of interest that are influenced by many factors, but it is unknown how precisely those factors actually are distributed in relation to the variable of interest in the population.

Importantly, this does not mean that quota sampling will always lead to problematic outcomes. Several researchers have demonstrated that indeed quota samples can be useful in particular contexts and if a lot of detailed work goes into the construction of the quotas and their interrelations as well as the design of the actual data collection (which we discuss in Chapter 3). Cumming (1990), for example, found, when comparing results from the administration of a survey through both quota and probability sampling, that differences between the two samples were either insignificant or substantially fairly small with regard to indicators on health promotion campaigns it engaged with. Vidal Díaz de Rada and Valentín Martínez Martín (2014) found that non-probability samples performed well or even slightly better than probability samples on socio-demographic indicators that quotas would typically be built around (e.g. as age and education), as those are specifically targeted. However, they also noted that probability samples achieved better results in secondary concepts, such as unemployment, which appears to be in line with the discussion presented above: when concepts cannot be easily attached to a small number of easily defined quota indicators, it becomes very hard to know whether the quotas chosen will actually result in a representative sample. This does not only apply to political attitudes but also other comprehensive concepts in the social sciences. Yang and Banamah (2014), for example, demonstrate the issues of using non-probability sampling in surveys engaging with social capital concepts. So while quota sampling can be useful, a lot of caution needs to be used in its administration and careful consideration applied to the specific topic and context regarding its suitability for such an approach that is prone to more biases than probability-based techniques.

Common alternative non-probability sampling methods

While quota sampling is indeed utilised quite extensively in survey work, other forms of non-probability sampling commonly are not, as they do not tend to aim at obtaining representative samples in the same way quota samples do. Most of these techniques are used more commonly for other research methods, where representativeness in a strict sense is less of a concern. However, it is worth briefly reviewing them, partially to also understand the contrast between them and the more suitable approaches we discussed above.

The simplest approach we could consider is called a *convenience sampling*. It is very easy to undertake and straightforward with a focus on quick delivery and, as the name suggests, convenience. The most illustrative image that this approach invokes is that of the interviewer standing on some street with a clipboard in their hand, simply interviewing whichever person comes up next to them, before moving on to the next person that comes along. Unless, of course, the population is that of all people who walk along that street specifically, the data collected from such a process would permit for very little in terms of generalisability or representativeness, as the sort of people who walk down a particular street will have certain distinct characteristics (e.g. they may work or live specifically in that area). There are analogous forms online, where websites may ask users on that site to fill in a survey. This might be a suitable idea for a company that wants to find out something about the actual users of the website, but if those questions were, for example, on political attitudes, while easy to collect, there would surely be a bias of what users would even look at that particular website, and therefore, we could not generalise to any meaningful population. Convenience samples therefore are rarely suitable for proper survey research; effectively they simply include the respondents a researcher can get easily.

A slightly more targeted approach would be *purposive sampling*. In this case, the researcher would have a clear target group or population in mind but would recruit respondents without the use of a probability or quota-based approach. This could often be the case, because there is no explicit sampling frame that could be utilised or the group under consideration is fairly narrow (e.g. executives in a particular type of industry in a certain region). The approach may be chosen, because researchers do not plan to generalise beyond the sample of individuals surveyed. Indeed, the respondents surveyed may be the target population, if it is small. In its own right, the findings based on such samples only have a rather narrow scope. However, they could be meaningful if, for example, used in conjunction with qualitative interviews of the individuals surveyed to obtain additional information from them or to gain basic data on a larger sample of individuals within the target group from whom interviewees are selected based on certain criteria that information was gathered for in the survey.

Another approach that is commonly used for small, narrowly defined groups is *snowball sampling*. This is a way to undertake sampling that is often used when it is very difficult to reach subjects, for example, because they are part of a small and vulnerable group or because they undertake illegal activities. Snowball sampling implies that the researcher will utilise their respondents in order to gain further respondents. The assumption is that those who fulfil certain characteristics that make them part of the target group for the sample might be able to generate contacts with others who also fulfil those characteristics. So, the sample grows continuously with the help of the research participants. Obviously, the nature of this approach means that, again, we are not able to generalise from the findings in this sample to any larger population in a statistical sense. Also, typically snowball samples tend to be rather smaller, which is why the approach is more commonly used in qualitative methods.

Respondent-driven sampling

Because of the limitations posed by non-probability methods that are not based on quota designs, surveying hard-to-reach groups in a way that allows us to make meaningful statements with the aim to generalise can be very difficult and is often seen as prohibitively difficult. However, researchers have developed techniques that aim to overcome some of these problems. In particular, respondent-driven sampling (RDS), initially developed by Heckathorn (1997), warrants some discussion. Typical snowball sampling, as discussed above, is not suitable for survey research at a larger scale if the aim is to ultimately make statements about a specific target population rather than just the group of specific interviewees. The two key problems encountered are the biases emerging from the initial selection of informants (usually a non-probability purposive or convenience sample) and their referrals to other potential participants. In hard-to-reach groups, there may be significant reluctance to pass on the details of contacts to a researcher.

Heckathorn (1997) developed an approach to address these issues practically. RDS begins similar to other non-probability approaches with a relatively small selection of initial respondents ('seeds') who are recruited on a convenience basis. They are offered some financial compensation for their participation in the survey interview. Instead of then simply asking them about other possible participations though, they are then asked to actively help recruit further participants and they are rewarded for those efforts additionally. This dual approach to the sampling and the focussed enumeration are key to RDS approaches (and reflect some of the suggestions raised by Rachel Ormston earlier in this chapter). Initial respondents are typically given a limited number of 'coupons' they are asked to pass on to other potential respondents inviting them to take part in an interview as well. When those respondents then take part in the interview themselves, not only do they receive a financial reward, but also the person who gave them the coupon does. So participants are compensated both for their participation in the interview and successful further referrals. Each new participant is presented with the same model and therefore recruits further potential participants. That way the bias inherent in the original selection of seed respondents is reduced, as the networks in the target population are broadened and the connections between initial and final participants become more remote.

Using RDS, Salganik and Heckathorn (2004) develop techniques for how to calculate proper estimates about the target population in a meaningful way. Furthermore, they show that under certain conditions the estimates are unbiased regardless of how the initial respondents were selected. However, while RDS provides an important methodological innovation, the conditions made have been shown to be rather strong in terms of their potential impact on the results. Gile and Handcock (2010) show that the positive evaluations of the researchers who developed RDS heavily depend on the applicability of the assumptions made in their models and that those assumptions often are not realistic. Nevertheless, they consider the approach important and useful but suggest that it could be developed further to reduce existing biases. In particular, they emphasise that the selection or 'seeds' should be carefully designed and the behaviour of further recruitment be carefully monitored and adjusted through a significant number of waves of recruitment (for which they make specific suggestions for the implementation). So while RDS requires a lot of careful considerations to be addressed, it can provide a potential avenue for survey researchers who want to overcome problems of studying hard-to-reach groups for whom classic probability or quota designs would not be feasible.

Box 2.3: Ask an Expert

John Curtice | Surveys, polls and exit polls

We often hear about surveys and polls. How would you distinguish between the two?

Well in truth, the distinction between a poll and a survey is partly in the eyes of the beholder. They're both attempts to try to interview a sample of people with a view to getting them to represent a wider population. A poll will usually tend to have, certainly in the UK, two characteristics that would distinguish it from what we might call a survey. The first is the sampling design. The word survey is usually applied to projects that in some way or other can be regarded as using a probability-based sampling design. So in some way or other there is a known probability of each member of the relevant population to be included in the sample, and indeed for all or at least virtual members of the population that known probability is greater than zero. The second distinction I think I would make is between exercises which are conducted over a long period of time and which therefore are able to try to maximise the response rates. And indeed, there is something where response rates matter. So there is a preselected sample of at least addresses if not individuals which you're going to interview. And it's only those that you interview and that's the end of it. A poll in contrast tends to be done over a short period of time, and it isn't always the case that there is some predefined population of people you're going to contact. That is, for example, if a poll is done by telephone, a pollster will often keep on polling until they've eventually got the thousand-person sample of what it is they want to do. But in any event, it's usually done over a pretty short period of time. And so therefore, it tends to be confined to those people who are (a) willing and (b) available in a relatively short period of time. So those are probably the two crucial distinctions that one might make. But in truth, these are both animals of the same kind. And in part, academic survey researchers like to call what they do – surveys. And commercial pollsters are happy for what they're called to be polls but some people will indeed attempt to call what I would call a poll a survey. And maybe even occasionally vice versa.

On election nights, we encounter so-called exit polls conducted on the day. How do they work?

All exit polls are incredibly geographically clustered samples, because they have to be done at (a sample of) polling stations. The first crucial insight on which the exit poll

(Continued)

operates is to say, hang on, it may be true, and we know it's true that support for Conservative, Labour and Liberal Democrat, SNP whatever, varies dramatically from one constituency to another. But the change in party support doesn't vary as much. Therefore, any sample of polling stations, however selected, has a better chance of estimating the change in support than the level of support. So the exit poll in the UK doesn't attempt to estimate levels, it tends to estimate change. Now that still leaves you with a question, 'but hang on, how can you estimate change if you don't know what happened last time?' The answer is the exit poll is conducted, wherever possible, at the same polling stations as last time. We pray the polling station boundaries have not been changed and you have to adjust on occasion. And therefore, we . . . get 140 estimates (which is the number of polling stations we sample) of change for each party, well 10 in the case of SNP. Some other advantages of this approach are that if Conservative voters, for example, are less willing to talk to an exit poll than Labour voters it doesn't matter so long as it's constant between the two elections. If postal voters vote differently it doesn't matter so long as the difference is constant across both elections, and so on. So although there is bias, as it is constant this method works. So that's what we do, and the other crucial bit of the exercise is you're then modelling those 140 estimates looking to see if you can identify how the change in support varies according to the known characteristics of constituencies. And you're trying to do that in order to improve the predicted seat outcomes, which are not one/zero estimates. So if, for example, a model ends up saying, 'well we think Labour is going to get 45 and the Tories get 44', we don't simply say that the Labour party is going to win that seat. We see, 'well there's probably about a 52% chance that Labour will win that seat and about a 48% chance the Tories will win', and we sum the probabilities. Now that matters if you've got a skewed distribution. If you've got a situation where maybe one party has got a rather high number of seats which we think it's going to win by a small majority and the other party has got a rather small proportion [of such seats], it's almost undoubtedly the case if you go for a one/zero calculation you would overestimate the number of seats that the party that's got lots of 51, 52% chances of winning would actually win [compared to adding up the probabilities across seats]. And that, on occasion, has made a difference and that's the secret.

Chapter Summary

- Survey research is featured frequently in media outlets, employed by politicians to back up arguments and cited in scientific enquiries on a range of issues. We encounter surveys, polls and the peculiar exit polls on election nights (and John Curtice discusses the differences between those terms).
- To ensure that the results from these investigations are meaningful and representative of the groups they are meant to study, it is crucial that we can appraise the quality of the samples that underpin those surveys.
- To maximise representativeness, probability sampling techniques theoretically are the best approaches we could choose. However, their feasibility depends on the availability of good sampling frames and sufficient resources. While we can adapt perfect random

sampling techniques through cluster, multilevel and stratified sampling approaches to overcome some problems, there are situations in which administering them can be prohibitively difficult.

- Non-probability sampling techniques face several problems and difficulties, but in particular quota-sampling approaches can at times help us to gain meaningful insights indeed, and respondent-driven designs may help in studying hard-to-reach groups. However, they require very careful consideration and analyses, as the researcher has to make a wide range of choices about the sampling design that can have a very significant impact on the results.
- Fundamentally, any approach chosen should always be transparently described alongside the results of a study, so that the potential limitations can be understood well and readers can examine how substantial the scope of the investigation really is and what group of people may or may not be represented by those who have been included in a particular survey.

Further Reading

Arnab, R. (2017). *Survey sampling theory and applications*. Academic Press.
This is a text for anyone interested in more advanced sampling theory and more formal ways of engaging with it mathematically.

Squire, P. (1988). Why the 1936 *Literary Digest* poll failed. *Public Opinion Quarterly*, 52(1), 125–133. https://doi.org/10.1086/269085
This is an article that provides more details about the case study used in the chapter and discusses why the classic, non-probability-based poll in the *Literary Digest* got the 1936 US presidential election wrong, although it had such a large sample size.

Sturgis, P., Baker, N., Callegaro, M., Fisher, S., Green, J., Jennings, W., Kuha, J., Lauderdale, B., & Smith, P. (2016). *Report of the inquiry into the 2015 British general election opinion polls*. British Polling Council and Market Research Society.
The 2015 general election in the UK found many people surprised by the results, as the results on the election day differed substantially from many pre-election polls. This report discusses in detail and in a rather accessible manner what happened and acts as a good case study.

3

SAMPLING MODE: HOW WE ACTUALLY COLLECT THE DATA

Chapter Overview

Introduction 32

Data collection methods 32

Considering non-response bias 43

Adjustments after data collection: weighting 46

Further Reading 51

Introduction

As we have seen in Chapter 2, there are many different ways to conceptualise sampling theoretically, and those considerations are very important, as they have implications for how we can interpret the generalisability of the results, and in particular, whether we can use an inferential logic to calculate estimates of confidence (which is discussed further in Volume 3 on inference of this series). However, even when we have a set of sampling methods identified that is suitable and feasible for our respective projects, further decisions have to be taken regarding the practice of our sampling endeavour. There are multiple methods for the data collection, and each comes with particular advantages and disadvantages. It is important to note that data collection methods and sampling approaches are *not* synonymous to each other. Each approach to sampling can be operationalised, at least theoretically, through a range of different data collection methods. Some methods may be more or less common to certain sampling approaches; however, we should not make the mistake to conflate the two. When this happens, debates about what the best approach to sampling may be often end up with imprecise discussions. Among the biggest developments in data collection methods over the past decade, for example, are internet-based data collection techniques. It is an often heatedly discussed topic amongst psephologists, but it is not particularly useful if we lump together all approaches to online data collection, which can be done using probability-based, quota and also convenience sampling approaches – with substantial impact on the quality of the sample.

So to properly discuss the implementation of sampling, we need to engage with the interplay between the theoretical sampling approach and the practical data collection. We will base our discussions in this chapter around that interplay and reflect upon advantages and disadvantages of each. We will also discuss to what extent the mode of the data collection may affect outcomes and engage with the important question of how we deal with the fact that not all people who we ask to participate in a survey may be willing to take part. Issues of potential biases, because of non-coverage error and variable rates of survey participation by different groups, are one of the most important empirical aspects to engage with in order to ensure representativeness. After introducing a range of options for how we may treat missing data in a survey, we will look at one way that is nearly always used in surveys to account for certain biases in the sampling: weighting.

Data collection methods

The availability of data collection techniques has changed substantially over time. While in the early 20th century the options were mostly limited to face-to-face or mail-order approaches (De Leeuw, 2005), the spread of new communication technologies first enabled the use of telephone interviewing and most recently internet-based approaches. In this section, we will dedicate space to discuss the four currently most commonly used techniques (face-to-face, telephone, internet and mailing methods) and discussions about combining them. The section highlights key advantages and challenges to using different techniques, but it cannot review all potentially relevant dimensions that characterise the respective usefulness of particular data collection modes. A good comprehensive examination is, for example, provided by Albaum and Smith (2012).

Face-to-face

Face-to-face methods are traditionally often considered as the 'gold standard' (Byrne, 2017) in survey research. Without the use of any technology or intermediary medium, they are, of course, the only possible approach to obtaining a sample. For large-scale surveys, they lend themselves well to probability-based sampling methods, but they can similarly be applied easily to the least representative sampling approaches, such as convenience methods (think of the person with a clipboard on a street). Obviously, if a person has to physically be in the same place as the survey participant, it becomes clear quickly that face-to-face methods typically are the most resource-intensive data collection techniques. They require a substantial amount of time, not simply for the interviews but also to geographically move to the locations where the interviews take place. For surveys spanning large populations (e.g. people across a whole country), interviews are therefore typically conducted by teams of researchers who may only cover certain areas, but nevertheless, the data collection period can either be quite long or the costs increase substantially, if one wants to have a greater number of researchers collect data simultaneously.

Given those concerns, why are face-to-face approaches then often seen favourably? Three main aspects can be identified: personal interviewing (a) permits a more direct engagement with respondents, (b) allows for the administration of multiple instruments in the presentation of the survey permitting a great deal of question types and observations to be explored, and (c) can be utilised well with certain probability-based sampling techniques (Fowler, 2012, p. 80). The first point is probably fairly obvious: speaking with a person directly permits for more in-depth and longer engagement. Face-to-face survey interviews can easily last for an hour, as they can, for example, be conducted in the person's home and involve more direct interaction. The interviewer has more opportunities to observe and react to verbal and non-verbal cues of the respondent signalling their relative interest or understanding, which is much harder on the phone and cannot be controlled in mail-order or online formats well (which is why telephone and online surveys typically take less than 20 minutes to complete). Also, a personal interviewer can, importantly, create rapport with sensitive or hard-to-reach groups, for example, if there are certain cultural or language barriers.

Regarding the second aspect, the personal administration of an interview allows one to change the mode of presenting questions. Researchers can ask questions verbally, supplement them with showcards or use visual cues (which telephone interviewers cannot, of course, but mail or online surveys are able to do), for example. On top of this, respondents can be allowed to answer certain questions on paper or an electronic device (*computer-assisted self-administered interviewing*). This approach is particularly employed when asking respondents sensitive questions they may not wish to disclose in person (as the closeness of the personal interview could then act negatively as a deterrent to be truthful), but instead lets people answer certain questions on their own without the interviewer taking part. Research has shown that utilising such techniques increases the degree of respondents' willingness to disclose sensitive information (Tourangeau & Smith, 1996).

The final point raised above is that face-to-face approaches can be implemented using probability-based designs, even without a pre-existing sampling frame, when the target population is fairly broad and area-based. For common attitudes surveys, for example, we often look at the population in a certain geography (e.g. country) at a certain age (e.g. working age, 16–64 years or of voting age). Applying multistage sampling approaches (see Chapter 2 for details), face-to-face data collection modes enable the operationalisation well. For example, after breaking down the country into regional clusters, one could randomly and proportionately select smaller areas (and potentially smaller areas within), such as postcode levels. Within each sample of ultimately selected sub-geographies, interviewers can be physically placed to then follow what is called a 'random route' procedure (the case study below illustrates an example of this technique). Such procedures stipulate a process to first select defined households randomly (e.g. by prescribing that each fifth household along a street may be contacted) and then, based on the number and typically the ages of household members, a randomly selected person within the household. Interviewers are typically only allowed to talk to the specific person in the specific household that was contacted following the randomising procedure – meaning they, for example, have to return to the household at a later time, if nobody is at home, or have to arrange for a new date and time to meet, if the person is currently busy. The procedure therefore permits a strong implementation of the idea of randomisation in a multistage design; however, it also explains why it can be so time- and resource-intensive if implemented comprehensively.

Telephone

Telephone interviews became popular with the widespread penetration of landline connections and started to be used as one of the main options for market and opinion research work in the USA and Europe from the 1980s onwards (De Leeuw & Berzelak, 2016, p. 142). Telephone interviews are typically much less costly to implement than face-to-face ones (Nathan, 2001), because there is no need for researchers to travel to different locations. Instead, based at a call centre, many interviewers can simultaneously call households and conduct the interviews from a remote location. While training interviewers is important, of course, following a script for the questionnaire is easier, as all instructions and questions are read out verbally only. There are no non-verbal observations that need to be incorporated or alternative input modes required, which simplifies the process. However, because respondents only engage through the phone, such interviews are usually shorter (commonly 10–20 minutes) than face-to-face ones.

Telephone interviews can be used with a variety of sampling methods. Crucially, again, they are often employed because it is fairly straightforward to develop a probability sampling design for them; however, quota-sampling designs can also be applied. Also, the sampling frame for telephone interviewing needs to include phone numbers. If we have a list of phone numbers for all members of the target population (e.g. for all subscribers of a particular service or members of an association), we can employ a straightforward random sampling approach, without the need to travel across the country to actually meet the people in question.

Box 3.1: Case Study

Random route procedure in wave 5 of the German World Values Survey sampling

(Adapted from infas, 2006)

- The starting address is randomly assigned by the computer.
- From there the interviewer follows the random route protocol to identify the starting and then all subsequent households.
- The start household is identified as the first household in the first house on the left side (when facing the street at the starting address) – this household is not interviewed.
- Every fifth household in sequence now is a target household (not every fifth house, but household).
- There are specific rules for the following:

  o How to count along multiple households (flats) within one house

(Continued)

- o What happens when the end of a road is reached (e.g. in which direction to cross the road to continue)
- o Which person in each household should actually be talked to according to certain random characteristics (e.g. relative age position)
- o What to do in special cases that would interrupt the route (e.g. cul-de-sacs or roundabouts)

- Identified target households cannot be replaced (i.e. if respondents are not at home, interviewers would return to the exact household again later).

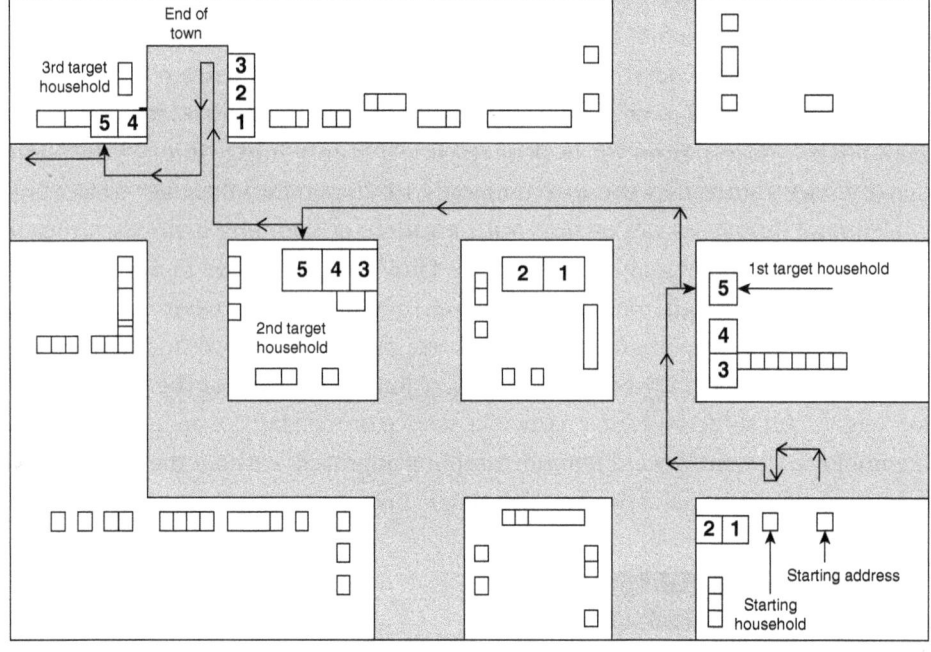

Telephone data collection techniques, however, are also valuable for larger-scale populations (e.g. all adults in the country). Phone numbers can either be drawn from national number registers, or, if those are not easily available, a commonly used technique that is employed is called *random digit dialling*. Commonly, computer-assisted, randomly generated numbers within the plausible set of digit combinations are generated and called. These procedures can of course also be applied in multiple stages, for example, by breaking down the country into regional clusters first and then calling respondents within each. Interviewers would begin a call by checking a respondent's eligibility through *screener questions*, such as establishing a person's age to make sure they are over 18 years, if it was a survey of people at voting age in a country, where that was the age of enfranchisement. Furthermore, and similar to face-to-face approaches, a randomisation of household members may be undertaken (e.g. to ask to speak to a randomly assigned age position within the household). In order for a proper randomisation to take place, efforts have to be undertaken to only engage with those respondents who were, respectively, selected in the contacted household. Similar to face-to-face interviews, interviewers would be expected to call the initial household back multiple times, if nobody picked up the phone or suggest to rearrange a different time for the call, or if the identified respondent was busy. Telephone-based data collection therefore does not imply fewer contacts with individuals than face-to-face methods, but it is differentiated mostly in the, often, simpler sampling frameworks based on phone numbers and the remoteness. While faster than face-to-face surveys, they still require a substantial amount of effort and personal contact time spent by interviewers, if aiming to develop a probability sample. A quota-based sampling design can speed up the process, but it results in a non-probability sample: in that case every respondent, who passes the criteria of initial screening questions, who an interviewer talked to would be incorporated into the sample, until saturation of respective quotas and cross-quotas was reached (see Chapter 2 for details).

Taking these considerations into account, telephone interviewing indeed seems like a very useful technique, with a substantial degree of flexibility to achieve meaningful samples, when verbal communication is sufficient to ask the questions from a survey questionnaire. However, the technique relies on a very important assumption. Unless we have a full register of phone numbers, both generated random digit dialling and the selection of numbers from broad telephone registers assumes that there is comprehensive phone coverage of the population. However, the penetration of landline telephones has decreased substantially in many countries (Mohorko et al., 2013), while registrations of mobile phones for phone books or similar lists have not compensated for it. Crucially, households without landlines differ in their composition from those with landlines and households who only have mobile phones tend to also have profiles different from both the other groups (Busse & Fuchs, 2011). When not being able to take this into account, decreasing rates of landline penetration could therefore create strong biases in telephone-based surveys and render them unusable in some contexts and have led to a reduced use of this data collection mode in survey research in many instances (Couper, 2017, p. 125).

Direct mail

Before the widespread use of telecommunication technologies, direct mail surveys were the main alternative to face-to-face administration, allowing for surveys to be conducted without an interviewer having to make contact with respondents in person. While the costs for direct mail surveys are typically higher nowadays than online surveys (Haskel, 2017), mainly because of fees for physically posting the questionnaire, they are significantly cheaper to administer than face-to-face surveys. Mail-order surveys can be undertaken through a variety of ways representing different sampling methods. Many initial considerations discussed above in relation to telephone surveys also apply here.

The researcher may have a list of addresses of all members of a particular organisation, for example, or a company may have a database of its clients. In such instances, using random sampling designs can be fairly straightforward. If additional information is available, for example, for patients of a particular treatment programme, invitations to take part in a survey could also be stratified according to such characteristics. Clustering and multistage designs often are feasible as well, when, for example, distinguishing different regional operations or sub-organisations, as addresses obviously point to particular geographical locations through postcode information (which could also be used for geographic stratification). If all we have is information about physical addresses through general address lists (e.g. based on official records or, if available, comprehensive electoral registers), we are able to engage with general population surveys.

Contrary to telephone interviewing, quota-sampling approaches are less meaningful for direct mail surveys, as it is not possible to stop a person from filling in the questionnaire and sending it back after asking initial screening questions. It is, of course, possible to filter out certain responses post hoc, when receiving the letters back, and instructions can be included to ask only for certain people to fill in the survey; however, with a direct mail survey, the researcher has less control over who participates in the first instance than in face-to-face or telephone-based interviewing. In particular, in contrast to face-to-face interviewing it is impossible for the researcher to know, even when post hoc screening questions are included, whether the person who filled in the survey really was the person they claimed to be. This suggests that unless we have specific address lists, for example, for subscribers to a particular service with information about possible respondents, direct mail surveys are more efficient for wider populations rather than very narrow ones, as many unusable returned questionnaires could be expected. It has been observed for a long time as well that response rates for direct mail surveys can vary tremendously, from less than 20% to as much as 80% (Fowler, 2012, p. 75), depending on a wide range of factors, including whether people were offered incentives to participate, who sponsored the survey (e.g. industry or government), how salient the topic was, what the target population was and whether people were followed up after the initial contact (Heberlein & Baumgartner, 1978, p. 456).

Crucially, direct mail surveys limit the set of possible respondents to those who are literate and specifically speak the language the questionnaire is printed in. Sometimes multiple formats can be sent out to people; however, that obviously increases the cost of the survey substantially and still does not overcome the requirement for literacy, which may exclude certain parts of the population, in particular in countries where literacy rates are not high. Similarly, contexts in which no comprehensive postal service and address systems exist are unsuitable for direct mail surveys. We find concerns about literacy and access also of importance in the next section, discussing internet-based data collection methods.

Internet based

Surveys administered over the internet are the most recent addition to the set of options researchers can choose from. They have grown in popularity a lot, as they permit to engage with participants remotely, but without a personal interviewer having to talk to each person, as they have to in telephone surveys, while also not having to wait for mail returns and, crucially, not facing the costs for postage. Online surveys therefore offer a data collection method that can often be quick and scaled substantially using technology and are often much cheaper than other methods of data collection. While, similar to paper-based direct mail surveys, they tend to be accessible only to those who are literate and, even more importantly, those with computer and internet access, which results in a range of biases (De Leeuw & Berzelak, 2016, p. 143), and they tend to offer substantial flexibility in terms of options for different ways of asking and presenting questions. Except for textual cues, images can easily be incorporated into questionnaires and even video or audio content used, with different options to create interactions with the person answering the questions, allowing for multiple input measurements that can all be recorded.

The quality of internet surveys depends substantially on the sampling methods chosen though. We come across many convenience samples online, when, for example, an online news site asks readers on the site to decide whether they agree with a legal decision that was made at the end of an article reporting on it. The responses to such a question will mainly tell you something about the composition of readers who look at such a website but will not provide you with much insight into anything that can be generalised to a wider population. Similarly, the online survey discussed in Chapter 1 of this volume, overestimating the amount of young people who had joined political parties in Scotland, was simply a convenience sample made up of respondents in school classes that already discussed political issues. Online surveys, however, can be administered through both quota-sampling techniques and forms based on probability sampling.

The majority of serious online surveys currently use non-probability sampling methods employing quotas. They either recruit respondents through existing websites or social media or, most commonly, rather than recruiting people anew each time, they are based on existing large-scale panels of potential respondents from which samples are drawn for a specific survey. The American Association for Public Opinion Research undertook an extensive study into the opportunities and challenges of using such panels (Baker et al., 2010), which highlights many of the relevant key issues one should consider. The main characteristic of such panels is that they are not representative of the target population per se. The recruitment is 'opt-in', meaning that individuals are recruited to and choose to take part in surveys for which invitations are sent to them, usually for some form of compensation for the participation (Couper, 2017, p. 128).

Consequentially, biases exist in relation to who is more likely to take part in such surveys. Also, the composition of the panels in the first instance affects the base from which people might be selected substantially, so it is crucial that researchers obtain as much information as possible about the initial panels that respondents are chosen from, which, however, can be difficult as commercial panel providers do not want to disclose too much information on their respective databases. However, the quality of the panel and its coverage of all relevant parts of the target population are crucial (Baker et al., 2010, p. 753). The recruitment process itself then tends to follow a quota-sampling design approach for which increasing complexity through cross-quotas can improve quality, but it may also extend the recruitment process and require the contacting of more individuals, as specific quotas become saturated, thus requiring greater resources. Overall, because of the limitations of opt-in approaches to the panel, certain biases are difficult to overcome and require substantial weighting (discussed below) often in reference to some existing probability-based survey or administrative data. However, even then the biases are often not overcome and remain larger than for probability sampling methods-based samples (Couper, 2017, p. 128). Panel participants may differ from non-participants, for example, in having 'less concern about their privacy, be[ing] more interested in expressing their opinions, be[ing] more technologically interested or experienced, or be[ing] more involved in the community or political issues' (Baker et al., 2010, pp. 746–747).

However, online panel providers have made great efforts to improve the quality of their panels and cumulatively aimed to expand and include more hard-to-reach groups, in particular. Sometimes, online panel-based samples have actually been better able to match, for example, actual results in voting choices in elections than probability samples (Baker et al., 2010, p. 713), because they could be administered more quickly and closer to an election. Indeed, as Couper (2008) shows, considering design choices for web surveys is of great importance and can affect their quality and procedural changes in many parts of the sampling procedure, including the recruitment process of respondents (Kaplowitz et al., 2012), which can enhance or diminish the quality of the sample. While many new innovations are being trialled to improve the quality of panels for non-probability web surveys and their implementation (Schonlau & Couper, 2017), Couper (2017) agrees with the American Association for Public Opinion Research Standards Committee (2010) assessment in relation to such surveys in not rejecting their usefulness but asking for a substantial degree of caution in the engagement:

> For academic researchers, nonprobability online samples are often the only affordable option. Such samples are not inherently incorrect, but they increase the risk of inferential errors over probability-based approaches and should be used with caution and with an explicit discussion of the likely inferential limits . . . (Couper, 2017, p. 130)

Internet surveys can, however, also be conducted using probability sampling approaches. This obviously is the case when we have a sampling frame that provides us with access to all members of a particular population who have email addresses or use a certain online platform that prompts can be sent from (e.g. workers within a particular organisation). For broader populations representing wider groups within society (e.g. all people aged 16 years or above in a country), we commonly do not have such sampling frames. What has been developed in some cases are therefore online panels, which, similar to the non-probability based ones used for quota-sampling within described above, provide a set of respondents who can be easily accessed for surveys. However, the crucial difference is that they have not self-selected into the survey but were recruited using traditional probability sampling methods – potentially entirely offline or using a combination of modes. An early and comprehensive example of such a panel is the Dutch LISS (Longitudinal Internet studies for the Social Sciences) Panel (CentERdata, 2018). The panel consists of 4500 households with 7000 individuals who have been recruited through a probability sample of households based on the population register of the Dutch statistical office, Statistics Netherlands. Households that did not have access to the internet or internet-enabled devices were provided with those to overcome internet access biases. Thus, the LISS Panel combines the convenience and flexibility of administrating surveys online with the quality of probability sampling techniques. Similar studies (Couper, 2017, p. 130) also exist in Germany (GIP [German Internet Panel]), France (ELIPSS [Enhancing Learning by Improving Process Skills in STEM]), Norway (Norwegian Citizen Panel) and the UK (NatCen probability panel), as well as the USA (GfK Knowledge Panel; USA). Such surveys still have some drawbacks worth considering. The quality of the initial recruitment determines the quality of the panel substantially, and response rates for inclusion can be relatively low, with biases in participation uptakes and response rates when being prompted to take part in a particular survey often lower than for other, more traditional, methods (Couper, 2017, p. 131).

Crucially, as with all other approaches, the choice for or against such a data collection mode is not trivial and comes with advantages and drawbacks that we need to consider. Importantly, surveys can be administered combining more than one data collection method as well, trying to harness the advantages of different approaches jointly, which we will discuss in the next section.

Combining methods

It can be of substantial advantage to combine multiple data collection methods for a survey to reduce different sources of error and types of bias which are more or less associated with particular interview modes. When reviewing different individual modes of data collection (De Leeuw, 2005, p. 246), none could be identified to work best in all circumstances and for all purposes. De Leeuw and Berzelak (2016, pp. 143–144) review the different modes individually, summarising key advantages and shortcomings: while face-to-face interviewing provides the greatest degree of flexibility and permits greater questionnaire length and complexity, the costs are very high and general response rates have been declining raising concerns of biases. Telephone surveys are less resource-intensive but still permit interviewer assistance through direct involvement; however, they are limited to verbal engagement, suffer from declining landline coverage and have low response rates. Direct mail surveys tend to have higher response rates than web surveys and can reach people without requirements for technology access, similar to face-to-face methods, but administration is costlier than online approaches and responses tend to be skewed towards older and more literate respondents who are not in transition. Online surveys can be very good value for money when using non-probability samples but can suffer from strong panel selection biases, require strong online access penetration and become more expensive to set up initially when probability sampling methods are desired.

There are two ways in which combined methods can be used: sequentially or concurrently. The former implies the utilisation of one message first, commonly to recruit people into a sample, and a second method to administer the survey. The latter would involve giving people the option to choose between different modes of responding to a survey. Obviously, the two can also be linked, meaning that people are recruited one way and then given multiple options to respond.

With regard to the first option, recruiting people using one approach and administering the survey using another, probability-based online panels are a good example. While online panels are great in terms of the practicability of administering a survey, they can suffer from strong selection biases, if they are opt-in, as discussed above. A recruitment strategy that does not rely on the internet in the first place is better suited to ensure that a panel can become representative. Therefore, selection into a panel could follow a probability approach, commonly either by inviting people to take part via mail or even face to face. Crucially, people without internet access can then still be reached, but then they have to be given the necessary hardware, software and skills to subsequently take part in the online panel (CentERdata, 2018). The other approach to combining methods, giving people the option to choose how to answer a survey, aims to minimise the biases that exist in single-mode designs. For example, people may be given the choice to answer a survey either online or via the telephone. While older people are less likely to have internet access, they tend to be more likely to have a registered landline in countries where landline penetration used to be high, while the opposite is true for younger people. Therefore, a combination of both approaches can be useful to minimise biases that are specific to one data collection mode. This can also be done stepwise: respondents may be given one option to participate (e.g. online), and for those who do not, a follow-up is arranged via telephone, mail or face to face.

Several studies suggest that, indeed, there can be positive effects of combining different data response modes (De Leeuw, 2005), such as increasing response rates and reducing particular biases inherent to single-mode designs (De Leeuw & Berzelak, 2016, p. 150). However, it is important to remain cautious nevertheless. The evidence is mixed whether errors are reduced in all combinations of mixing response modes and the specific configuration and implementation requires careful thinking (Couper, 2017, pp. 132–133). While some biases may be addressed, new sources of error can be introduced through combining response modes concurrently, because rather than only representing slightly different sets of respondents, the mode of interviewing itself may actually affect the way people answer questions. This in turn makes it more difficult to assess whether the responses from all modes of interviewing can be treated equivalently. People are more cautious in revealing sensitive information to a person interviewing them (Tourangeau & Smith, 1996), and more impersonal forms of data collection, such as mail surveys, tend to produce fewer socially desirable answers (De Leeuw, 2005, p. 246). Furthermore, differences may not only be found in terms of the types of responses but also in the structure of concepts that are meant to be analysed. De Leeuw et al. (1996), for example, found that the data collection method can have an impact on structural models in the analyses the survey data is used for. While some studies have shown results to be equivalent in certain circumstances, others have found substantial differences between modes (McDonald & Adam, 2003). Combining methods can be very helpful, but rather than assuming equivalence outright, we should always investigate potential differences empirically for the research we undertake.

Considering non-response bias

In Chapter 2 of this volume, we already discussed what problems can arise from missing or incomplete sampling frames. It may not be easy to draw a sample from the target population straightforwardly. If the frame is incomplete, we may experience non-coverage error, because, even if drawn at random, not every person in the target population would have an equal chance of being selected, thus undermining the probability-based approach. This is particularly problematic, if there are imbalances regarding which groups of the target population may or may not be covered. If there was a bias in that regard, this bias would eventually translate through to our sample, resulting in a corresponding bias. In addition to distortions based on incomplete *coverage*, we may also, however, encounter *non-response* biases. Even when we have a complete sampling frame and are able to apply random sampling methods to recruit respondents, we may encounter a new set of difficulties: not all selected respondents may be equally likely to take part – creating new distortions.

Individuals participate in social surveys voluntarily (except for census-style enquiries). Therefore, they may refuse to take part for a variety of reasons, ranging from simply not wishing to spend the time, not being interested in the incentive provided (if one is offered) or not trusting the organisation commissioning the survey to feeling uncomfortable to disclose information about the topic of the survey, because it may be a sensitive topic or addressing a taboo (Stoop, 2012, p. 128). Overall, response rates to surveys have fallen substantially in many (Kreuter, 2012), albeit not all, countries (Stoop et al., 2010). Lower response rates are of concern, especially when non-respondents are different from respondents in their characteristics. That situation would result in samples with biases, thus reducing their respective representativeness. However, response rates are not related perfectly to the degree of bias that may exist, so crucially, the focus should be on specifically investigating that potential bias caused by certain groups of people being more or less likely to take part in surveys (Stoop, 2016, p. 410).

Many factors influence the likelihood of whether certain individuals or households will participate in a survey or not. Groves and Couper (1998, p. 30) provide an analytical framework for survey cooperation. They distinguish between factors that are under the control of the researcher, in particular the design of the survey and the interviewer, and factors that are out of the control of the researcher, namely, the social environment and the household or individual within a household (see Table 3.1). As we already discussed, the design of the survey itself may affect participation, for example, not only through the mode of administration or the scope of the population that is to be studied but also through factors such as the length of the survey or the cognitive complexity it requires from respondents and the topic under consideration. However, interviewers can also influence the likelihood of participation, both based on socio-demographic characteristics (e.g. by building trust when being of a similar demographic as certain groups that could be described as vulnerable) and their skill and experience (which we will return to in Chapter 5). External conditions can affect

Table 3.1 Influences on survey participation according to Groves and Couper (1998, p. 30)

Under researcher control	Survey design	• Topic • Mode of administration • Respondent selection
	Interviewer	• Socio-demographic characteristics • Experience • Expectations
Outside researcher control	Social environment	• Economic conditions • Survey-taking climate • Neighbourhood characteristics
	Household(er)	• Household structure • Socio-demographic characteristics • Psychological predispositions

survey participation as well. Macro-conditions, such as economic situations, as well as the general attitude of populations to surveys may act as moderating influences and so can neighbourhood characteristics. At the household level, in addition to socio-demographic and household compositional differences, psychological predis-positions may also affect the likelihood of a person being willing to take part.

However, we need to be careful in our interpretation of particular factors identi-fied as being associated with a greater or lower survey participation tendency. Many different factors we observe (e.g. age, education or housing tenure) may be at least partially expressions of each other. Therefore, we should treat them as correlates only, unless we can examine the causal mechanisms precisely. Building on that impor-tant caution, Stoop (2012, pp. 126–128) identifies four main reasons why people may choose to participate in a survey: (1) they may enjoy taking part in a survey, because they like giving their opinions on particular issues, which is likely to favour the participation of people with higher cognitive skills and an interest in politics; (2) participants may be motivated if an incentive (monetary or non-monetary) is offered, which would likely lead to greater engagement of less well-off people; (3) the survey may be perceived as important, because the respondents' opinions count and there are likely improvements to, for example, health or living conditions that the results of the survey would support, which would likely see people using public assistance or who want to ventilate their opinions participate more; and (4) respondents may take part, because they consider that the survey participation could be important for oth-ers, such as society at large, science or a specific community of interest, which would see people who are religious, who are socially integrated, who undertake voluntary work or who have an interest in politics more likely to participate.

Researchers should consider these factors in advance of administering their surveys in order to reduce biases in non-response. While respondent cooperation rates have seen some declining trends, researchers have developed several meaningful responses that can be adopted to improve the quality of samples (see Glaser, 2012, for a detailed discussion). Additionally, after the collection of their data, they should also investi-gate to what extent biases may exist in terms of those selected, either in contrast to non-respondents (if data on them is available) or known characteristics of the target population (again, if those are available). When dissonances are identified in terms of over- or under-representation of certain groups, we are able to partially adjust for that through weighting procedures, which will be discussed in the next section. In addi-tion to examining non-response for whole cases, such as individuals or households, it is also common that not every survey participant will respond to each question. While the former, discussed here, is called unit-non-response, item-non-response is also an important aspect to address systematically in the analyses.

In particular, one approach to engaging with item- and unit-non-response issues are different approaches to imputing information. Rather than simply disregarding respondents with gaps in the data, probabilistic techniques are used to estimate the most likely response for the respondents in question given all the information we have about them and the patterns identified in the rest of the sample (or prior information from other studies on similar populations). The best approach to the implementation of such techniques for dealing with missing data is an area of extensive study and debate within survey research, and researchers should carefully consider different options of engagement before starting analyses.

Adjustments after data collection: weighting

Nearly all surveys suffer from some biases in terms of the respondents. For example, surveys on political issues tend to have a greater proportion of people who take part in elections than actually do in practice. When the discrepancies are too large, a sample may be inadequate for research. However, small deviations can be accounted for meaningfully using weights. Weighting is a standard process that tends to be applied for all social surveys and can be undertaken at varying degrees of complexity (Lavallée & Beaumont, 2016). While the practice can require a fair deal of mathematical operations and careful considerations (for a comprehensive summary see, e.g., Dorofeev & Grant, 2006, pp. 45–78), the basic principle is rather straightforward. With regard to the concerns raised here, our main aim for using weights would be to make our sample 'behave' as if it was more similar to known population parameters, thereby *calibrating* the sample to correct the deviations in the sample statistics compared to the associated population parameters.

Applying weights to account for response biases

Let us revisit the example from Chapter 2 of a hypothetical survey conducted in Poland (Table 3.2). We know the population distribution for sex and age from national statistics. Therefore, we can calculate what distribution of the sample (with 1000 participants) we would expect if that distribution was perfectly represented. If that was the case, the weight for each case in relation to these two variables would be 1.0 – we would not need to make any adjustments. However, as we discussed above, there are nearly always going to be differences in response rates for different groups of the population. Here they are not very large, but present. For example, we would expect 480 male respondents, but only 450 took part, while instead of 520 expected female participants, we had 550. The men are therefore slightly under-represented and the women over-represented. We can adjust for this by giving each male respondent a slightly greater weight in our analyses and correspondingly give a lower weight to each woman. The weight is calculated by dividing the expected distribution figure by the actual distribution figure. So, for example, for women we would divide 520 by 550, which is 0.945, which would be the weight applied to our analyses for each woman, giving them in the aggregate the same proportionate impact on the estimations as their proportion of the target population would suggest. Obviously, each case has multiple characteristics, such as sex and age. In our sample, 18- to 25-year-olds were under-represented. Their weight would be calculated as 130/100 = 1.3. So for a 22-year-old woman, her overall weight, combining both weights, would be 0.945 × 1.3 = 1.229.

Table 3.2 Calculating weights for a survey sample from Poland with reference to population statistics (Eurostat, 2016)

Demographic Group	National population distribution (%)	Expected distribution in sample (Total = 1000)	Weight factor if expected distribution was achieved	Actual distribution in sample (Total = 1000)	Associated weight factor based on actual distribution
Male	48	480	1.0	450	1.066
Female	52	520	1.0	550	0.945
Age (years)					
18–25	13	130	1.0	100	1.300
26–34	18	180	1.0	170	1.059
35–44	17	170	1.0	200	0.850
45–54	16	160	1.0	180	0.888
55–64	18	180	1.0	180	1.000
65+	18	180	1.0	170	1.059

Weighting procedures employ a higher degree of sophistication and can take further relationships between variables into account, but the basic principle is the same in terms of making adjustments between the actual sample and the desirable distribution. This is discussed further in Volume 5 on secondary data analysis of this series. It is important to be cautious about applying weights though. What we effectively are doing is to assume that the respondents in each group in our sample, on average, behave like the members of that same group in the population on average. In other words, we assume that the 18- to 25-year-olds, who are under-sampled, are not prone to a sampling bias in some other way that we have not accounted for through other weights (e.g. they might be young people who are more interested in politics). Weighting cannot account for dissonances that we are unable to estimate or observe. That is why, in particular for samples that apply very large weights, we should be rather cautious about the use. Even if we are able to match the demographic parameters in the population with our sample, we may not be accounting for some of the selection biases leading to differential non-response that we discussed earlier. So even when through weighting we achieve an effective sampling distribution in our analyses that matches the socio-demographic profile of the population, we cannot automatically assume that it is representative with regard to other variables without caution.

Applying weights to account for attrition

Above we looked at response biases in cross-sectional surveys. Effectively, we were looking at deviations in the sample characteristics compared to the expected distributions of those characteristics in the target population. However, survey research often is interested in studying change, so a panel of respondents might be interviewed multiple times over several weeks, months or even years. Even when we have a perfect random sample of our target population initially, there may be non-response biases affecting our sample as we progress through multiple waves of data collection over time. Some respondents may not be available for one round of interviews but reappear at a later stage again, while others may stop participating altogether (e.g. because they choose to do, move or pass away). The issue of **attrition** (respondents stopping to take part in further waves of data collection) is one of the most important ones in assessing data quality in longitudinal research using panels.

Attrition is less of a concern for the quality of the data, if all groups of the respective target population are equally likely to discontinue their participation in the survey. If everyone has the same probability of continuing/not continuing, we would not expect to find any biases overall in terms of the results. Similar to the cross-sectional analyses discussed above, however, often some people are more or less likely to continue to participate, which can end up resulting in serious biases the more waves are being conducted. So careful analyses of attrition patterns are important to identify potential problems. When these are correctly noted, adjustments can be undertaken using weights to address attrition biases. As many studies have shown, attrition is typically very unevenly distributed in a sample. The Michigan Panel Study of Income Dynamics, for example, saw roughly half of its respondents drop out between its inception in 1968 and 1989, but some groups were much more likely to do so than others (Fitzgerald et al., 1998). In this case, participants with lower socio-economic status and respondents with more unstable marriage, migration and earnings history had higher rates of attrition, though the authors found the effect on estimates to be limited overall.

The application of weights therefore is not always a straightforward decision. The goal is to compensate for attrition biases, but how well that is done depends on the quality of the weights and the assumptions made when computing them. In their study of the European Community Household Panel, Vandecasteele and Debels (2007) show that applying longitudinal weights can reduce biases, but in some instances, the application of weights actually made things worse, as the originally developed weights did not properly capture the likelihood of attrition sufficiently well. They demonstrate that the quality of weights can be improved, however, when not only demographic characteristics but also factors, such as covariates, that relate to the interview process itself are taken into account. In short, the application of weights is commonplace and potentially useful in reducing attrition biases, but the use of weights does not necessarily mean that the quality of results is actually improved. This depends on how good weights are at capturing the biases inherent in non-response in the first place – which is why it is important to carefully understand and assess the design of weights that are used in survey research.

Box 3.2: Ask an Expert

John Curtice | Mistakes in UK general election polling

What issues do we need to pay particular attention to in online election polling?

One thing that's pretty clear from the fairly substantial literature of this . . . is that basically this is not an issue of mode. In other words, there's nothing inherently wrong with interviewing, if that's the right word in this context, people over the internet. And that it is no better or no worse as a mode of interviewing than telephone or face to face. The issue is an issue of sampling. In practice, no survey manages to make contact with all elements of the sample that they have chosen. And therefore, there is a risk that those who respond to a survey are atypical of those who do not. Even two probability surveys can vary in the current sample they get . . . Internet polling in the UK is [mostly] done amongst people who have been recruited into signing up and saying, 'yeah I am willing to be contacted to do your surveys'. But it's not done amongst a random sample. It's only amongst people who've agreed to sign up. Now the risk therefore with that kind of exercise is that the kind of person who is willing to say, 'yeah I'm willing to be contacted to do a survey and maybe I'm going to get paid 50p or £1 a time', is not typical. We have in election polling people that are more likely to be people who would go out and vote which you might think is fine because this is what we want to do – to ascertain in the election poll how people are going to vote. Except that one of the things that we've learned and one of the major lessons in the polling industry in the last 2 or 3 years is that actually because turnout varies by demographics and at the same time correlates with vote choice. If you overestimate, for example, the probability that young people would go to the polls, because young people are also more likely to vote Labour these days, you are therefore at risk of overestimating the Labour vote.

(Continued)

Is there a way to evaluate the quality of an election poll well? What are the things you look at?

Good question. You look at how heavily has this poll had to be weighted. The more a poll has to be weighted, the more suspicious you get partly because actually in practice that therefore reduces the effective sample size. So you may have interviewed 1000 people, but if the weights attached to some respondents are, you know, weights of two or three, the effective sample of size is well below 1000 and therefore the sampling error is greater. You're particularly interested in how much it had to be weighted on key variables. And if you're talking about a political poll, the key variable is how people voted last time. You're certainly looking to see how much that had to be weighted. You're then also looking to see has it been weighted correctly. So, for example, one of the things I do when I have 5 minutes to sit down and go through polls is actually go through them and say, 'well, you know what, what if they weighted their social grade to their age distribution etcetera'. So, just looking to see whether they look right or not. One of the reasons why I was getting worried about some of the polls in 2017 was one of the things that we knew from high-quality data, such as the British Social Attitudes Survey, is that basically younger people were less likely to turn out and vote in 2015. Although 18- to 24-year-olds constituted a low percentage of the adult population in the UK, they actually constituted 9% of those who turned out to vote. Therefore, one of the things that perhaps made sense to do is to weight the age distribution according to the proportion of people who voted in each age group [in 2015], so that they constituted the actual voting population as opposed to the electorate. That obviously comes with a risk, because these things may not be constant over time. [Pollsters] ended up with weighting young people not according to their reported probability of voting but what they thought was the probability of people voting last time. You were ending up with estimates of vote choice based on samples where only 4% or 5% of the voting population was expected to consist of 18- to 24-year-olds. They basically did end up with an awful lot of double weighting. And . . . that kind of made me go, hang on, I know what you're trying to do there but you've over-egged the pudding.

Chapter Summary

- In addition to considering the sampling technique, we need to carefully think about how we go about collecting the data. Different modes of data collection can be identified. Which one we should choose depends on a wide range of factors ranging from the availability of resources to important consideration of mode effects that can, but do not have to, occur.
- Considering them is particularly important when we engage with sensitive topics, where interviewer effects could affect the way people respond.
- Crucially, we should not conflate sampling method and data collection modes. Internet-administered surveys are often discussed as if they were one homogeneous group of surveys, but critiques often specifically focus on non-probability-based ones.

- However, both probability-based internet samples exist in multiple countries, too, so we should distinguish between the sampling technique and the specific operationalisation through a data collection mode.
- Different modes can also be combined, but careful analysis about potential mode differences should then be undertaken. Regardless of the specific mode, researchers should analyse the impact of differential participation by various groups of the target population and consider the use of weights to adjust for biases.
- However, once again, we need to be cautious about the precise application. Even when calculated correctly in a mathematical sense, weights can distort results, when unobserved parameters are not adequately accounted for. This can affect results substantially, as the discussion of polling in UK general elections demonstrates.

Further Reading

Couper, M. (2017). New developments in survey data collection. *Annual Review of Sociology, 43*, 121–145. https://doi.org/10.1146/annurev-soc-060116-053613
This is a very comprehensive article that discusses how survey data collection has developed, what the advantages and disadvantages of certain modes (in particular online data collection) are and whether combining methods can be useful.

De Leeuw, E., & Berzelak, N. (2016). Survey mode or survey modes? In C. Wolf, D. Joye, T. Smith, and Y. Fu (Eds.), *The SAGE handbook of survey methodology*. Sage.
This is a great piece discussing how and when it can be useful to combine multiple methods of data collection within the same survey and what one should look out for in doing so.

Fowler, F. (2012). *Applied social research methods: Survey research methods* (4th ed.). Sage.
This book is a comprehensive account of survey research methods and their practical applications that would provide readers with in-depth insights to further the points raised here.

4

QUESTIONNAIRE DESIGN: ASKING THE RIGHT QUESTIONS

Chapter Overview

Introduction 54

Distinguishing between different types of questions 55

What makes good questions – and some common pitfalls 62

Thinking about the questionnaire structurally 72

Conclusion 78

Further Reading 79

Introduction

So far in this volume we have discussed how to make sure that the samples we obtain actually provide the quality we need in order to make statements about the respective population we are interested in. When our sample is flawed, we might be asking interesting questions, but the analyses we then conduct would not permit us to make strong inferences. So spending a lot of time on getting the sampling right is important. However, while this is clearly necessary, it is not sufficient to only focus on that aspect of a survey. Sampling effectively provides us with the structure of our survey; it forces us to clearly define our population and develop a strategy for how to recruit participants who are as representative as possible of that population. When we achieve a good outcome in that regard, we have provided a strong foundation for our survey. Having said that, at that point it is still not filled with any content, of course. In order to collect data, we need to ask respondents questions, and their responses will ultimately become the values in our data set. So the quality of these questions is crucial to make sure that we are actually measuring the things we are trying to operationalise.

This is even more pertinent considering that space on questionnaires is usually rather limited. Especially when the survey is administered over the phone or internet, 10–20 minutes of time is often the most one could expect to use, which is a fair amount for a short poll but not very much for a comprehensive survey. It is therefore very important to plan well and to ensure that the questionnaire development is done with a close orientation to the research questions one aims to answer (Jann & Hinz, 2016). We need to make sure that the sorts of questions we ask correspond well to the analyses we ultimately want to undertake. This includes considering not only what questions we need to ask about the core issue of the investigation but also importantly, and frequently not thought about enough, the questions that we may need to relate the core issue to, in order to understand for whom and why they may look one way rather than another.

We begin this chapter by providing an overview of the common types of questions one could ask and then discuss a wide range of common pitfalls that can occur in designing questions, to make sure we avoid the mistakes and develop good-quality instruments in our survey. After focussing on individual questions, we then widen our scope to think about questions as groups of items and engage with the important issue of question order and see how it can affect results substantially. Furthermore, we will specifically discuss why it is important to think explicitly about how our questions should be linked together to provide meaningful analyses and good degrees of validity and reliability. Before fully launching a survey, one should always conduct some piloting to find out how well the designed questions work. We will discuss how to do that in more detail in Chapter 5.

Distinguishing between different types of questions

Open- and closed-ended questions

Statistical analyses require a data set in which we are able to code the information we obtain from a survey in a quantitative way. For each case (e.g. an individual respondent and household) and each variable (e.g. age or educational attainment), we need to be able to associate one discrete value. That value can be given meaning, of course, so that, for example, one numerical value ('0') is coded as one age group and another ('1') would stand for a different, but also clearly defined, age group. However, this fundamentally requires that the question provides us with distinctive categories that we can code subsequently. Questions that enable us to do this straightforwardly are called **closed-ended questions**. This means that those questions have a distinctive set of discrete answer options that the respondent has to choose from in a predefined way (we describe the main types below). The opposite of this are **open-ended questions**. In quantitative research, these are less common. Open-ended questions permit the respondent to formulate their own response without a predetermined format, usually in textual form. The problem is that this means that each answer initially is unique and can therefore not be straightforwardly coded numerically. Also, open-ended questions in surveys require much more time from respondents to answer and are therefore frequently skipped (which may create biases in terms of which people are actually responding and who is not). Open-ended questions are therefore more common in qualitative research; however, this does not mean that they cannot be utilised in surveys. Sometimes, the information is collected and analysed using qualitative methods but is linked to the quantitative results of the survey. It is also possible to apply a post hoc coding to the open-ended questions by going through all the data entries. However, especially for large samples, that is very onerous. Finally, quantitative techniques such as text mining and word-based cluster analyses can be used to analyse free-text comments when that is the explicit goal of the research project. However, as stated above, the main part of most large-scale surveys consists of closed-ended questions, the answers of which can be directly translated into discrete values. We will look at the various manifestations thereof in turn below.

Single-answer multiple-choice questions

The simplest variant of a closed-ended question are those that provide multiple answer options (multiple choice) from which the respondent is asked to only select one. This could be a binary choice (e.g. 'yes' or 'no'), or it could be from a longer list of options. Such a question can be coded very straightforwardly with one variable corresponding to the question and each answer option being one of the possible values that could be selected. An example for such a question would be as follows:

Which political party, if any at all, do you feel closest to?

1. I do not feel close to any political party.
2. Conservative Party
3. Labour Party
4. Liberal Democrats
5. UK Independence Party
6. Green Party
7. Other

The question clearly asks for one discrete answer, requiring the respondent to think which party they have the greatest affinity for. The superlative ('closest') is important here. If the question had asked which party a respondent felt 'close to', we would have to permit them to select multiple options. Note the first answer option, 'I do not feel close to any political party'. Including this means that we are able to capture respondents within this one question if they feel no affinity to any political party. If we did not have this option, a prior screening question would have been required to establish whether people felt close to any party at all – for example,

Of all the political parties that exist in this country, would you say that you felt close to at least one of them, or would you say that you did not feel close to any party?

1. I feel close to at least one political party.
2. I do not feel close to any political party.

Only respondents who had chosen the first option would have then been asked the previous question (but consequentially with the first answer option removed) in this case. We should note that the list does not take into account *all* existing political parties. That is to prevent the answer option lists from becoming too lengthy (which may be a particular concern in telephone-administered interviews, where reading out long lists may deter respondents from continuing with the interview). The 'other' option nevertheless permits people to say that they feel an affinity to a different party. However, if an option is explicitly listed, the likelihood that it is chosen tends to be higher because it is easier to consider that option amongst the others listed. So such choices should be made carefully. It is also common in cases like this to add a short open-text option attached to 'other'. Consider the following:

Which of the following religious affiliations would you say you'd belong to, if any at all?

1. I do not have any religious affiliation.
2. Catholic

3. Protestant
4. Islam
5. Judaism
6. Other (Please write in your affiliation.)

The write-in to the 'Other' option would obviously be open-ended and cannot be coded directly. Post hoc coding would be required if, for example, we found that another denomination was selected very often and we would like to be able to conduct analyses specifically for that group. We would therefore need two variables in our data set to record this information: the first one would note which of the six options a respondent has chosen, and the second one would record the written entry for those respondents who had selected option 6, 'Other' (while for the other respondents, there would be no entry for the second variable). Providing a write-in possibility for the 'Other' option can serve an additional function: the respondent is given the feeling that their personal position, even if not listed explicitly, is being recognised, which is affirming and may prevent negative feelings about the survey.

Multiple-answer-item questions

When discrete answer options are presented, they have to be distinct, but it may not be problematic to consider that several of them could be selected validly at the same time. It depends very much on the type of question asked. Typically, questions that permit multiple answers to be selected at the same time present some sort of list of options, as in this example:

Please say which, if any, of the following words describe(s) the way you think of yourself. Please choose as many or as few as apply.

1. British
2. English
3. European
4. Scottish
5. Other (Write in)
6. None of these

Questions such as this one emphasise that the concepts they aim to capture may be multidimensional and permit respondents, as in this case, to express that their national identity might be an interplay of affiliations at several levels. Importantly, we need more variables in the data set to then code the responses to such a question. For each of the options listed, we would have to create a variable for which we would record whether the option has been chosen or not. The number of answer options that the researcher permits the respondent to select can also be restricted. Such questions are usually used in simple prioritisation exercises, where respondents are asked to think about issues with the highest salience in a defined regard but are permitted to select more than one option, without explicitly deciding on a ranking between the selected items, such as in the following example:

> When you decide on the location for a vacation, what, to you, are the most important issues that you consider when making your decision from the list presented below? Please select at least one and at most three items from the list.
>
> 1. The quality of the accommodation offered
> 2. The distance of the destination from your home
> 3. The weather at the destination
> 4. The costs of travelling to and staying at the destination
> 5. The costs of food and drink at the destination
> 6. The availability of activities for children at the destination
> 7. The language spoken by most people at the destination
> 8. The type of food typically served at the destination
> 9. The quality of the nightlife at the destination
> 10. The variety of cultural offerings at the destination
> 11. Other (Write in)

Again, we would need to create one variable for each item in our data set to record whether it has been selected or not. Reading the question above, we should note that it implies that a screening process had taken place. There is an implicit assumption that the respondent goes away on vacation, as there is no option to say that they do not. It may be that a screening question was asked beforehand or that we know from the context that they do – if, for example, the question was asked as part of a survey by a travel company the respondent had recently completed a trip with. We will discuss the implications of ensuring that the answer options are comprehensive further below.

Likert scale questions

Very often in survey research, we are not simply interested in respondents selecting one discrete option from a presented list but want to establish, in a nuanced way, to what degree they support or oppose a particular perspective. Likert scale questions are one of the most commonly used instruments that permit us to operationalise precisely that idea. They are commonly 5- or 7-point scales with specific verbal attributes attached to each position on the scale. Those verbal attributes are symmetric, presenting a balanced one-dimensional scale with opposing positions at either end. Consider the following example:

To what extent do you agree or disagree with the following statement? 'Sometimes politics and government seem so complicated that a person like me cannot really make sense of it all'.

1. Strongly agree
2. Agree
3. Neither agree nor disagree
4. Disagree
5. Strongly disagree

The scale options are organised so that each equivalent response on either side of the scale directly corresponds to the other one to maintain the symmetry. The question invites people before formulating the statement to agree or disagree, which is important so as to not present one direction as the starting point (which would otherwise be leading – as discussed further below). This question provided a neutral midpoint as an option for respondents to select. However, Likert scales can also be constructed without such a midpoint, such as in this case:

On the whole, are you very satisfied, rather satisfied, rather dissatisfied or very dissatisfied with the way democracy is developing in your country?

1. Very satisfied
2. Rather satisfied
3. Rather dissatisfied
4. Very dissatisfied

Not having a neutral midpoint forces respondents to indicate in which direction they are leaning. People cannot 'sit on the fence' but have to decide what perspective comes closer to their own views. There have been extensive debates in the literature on survey design discussing the question of whether it is generally better to include a neutral midpoint option or not. Not including it can, for example, make it easier for people to indicate a view they do hold but feel slightly uncomfortable revealing, as a midpoint option may be a way of masking their actual opinion on an issue (Garland, 1991). The fundamental question is whether the midpoint reflects a genuine position of preferring neither one option nor the other or whether it is effectively a marker of indecisiveness (Raajmakers et al., 2000), as this would strongly affect the interpretation of such a midpoint. Some researchers suggest, however, that, while there is an influence, the actual impact of potentially differing understandings of a midpoint option may be small in the aggregate (Armstrong, 1987). Crucially, the choice of including or excluding a midpoint is not trivial, but the decision may depend on both the context and the research aim. Both approaches can be justified, but the important point is that the interpretation has to be adjusted accordingly. If no midpoint is offered, we have to be aware that some people may not actually have a preference between the two sides of the scale but will provide an answer if forced to do so. We should then interpret their response as such, a selection from a limited set of options they were provided with. When we do include a midpoint, we should consider that its selection could reflect either a genuine non-preference for either side or a degree of indecision. In order to account for this, Likert scales are sometimes complemented with an option that is off the scale and tries to explicitly capture the responses of those respondents who simply cannot make a choice. Consider again the first example presented – regarding a self-evaluation of political understanding. If a sixth option was added that explicitly said 'Don't know', respondents would be given a clear option to say that they did not feel able to make up their mind in favour of any of the options provided, including the explicitly neutral one. We will discuss the importance of such 'Don't know'-type responses further below. The coding of Likert scale questions is straightforward. As only one option can be selected, only one variable needs to be used in the data set to record the respective response.

Numerical scale questions

There are other forms of scale-based questions that are utilised in surveys. Commonly, scales that do not follow a Likert-style approach, containing explicit verbal attributes for each value, are employed when respondents are asked to select their position on a wider spectrum of choices and where the researcher is interested in evaluating more nuances between different survey participants. Numerical scales are often developed on the base of 10- or 11-point scales but can be based on other choices. Usually, the participant is clearly informed about the meaning of the end points, which should be labelled clearly. Similar to Likert scales, the labels associated with the end points should be clear opposites at the end of a possible spectrum of answers and equivalent in their degree of distance from the middle. The same issues discussed about neutral midpoints for Likert scales also apply here. They can be included or not, and interpretations should be adapted accordingly. When a neutral midpoint is included (e.g. on an 11-point scale from 0 to 10), the midpoint is sometimes labelled as well to explicitly indicate its role. Longer numerical scales are not that straightforward to administer over the phone, as, without the specific labels for values read out for a Likert scale, it can be quite abstract to consider such a scale without any visual stimulus. In other forms of data collection modes, it is common to actually visualise the scale explicitly, on a showcard or screen. In surveys administered online, often respondents can actively move a slider alongside the scale to increase the engagement with the finding of a position along the scale. Here is an example of an 11-point scale with a labelled midpoint (Table 4.1):

Taking all things together, how satisfied would you say you are with your life nowadays on a scale from 0 to 10, where 0 means totally dissatisfied and 10 means totally satisfied?

Table 4.1 Example of an 11-point scale with a labelled midpoint

0	1	2	3	4	5	6	7	8	9	10
Totally Dissatisfied					Neither Satisfied Nor Dissatisfied					Totally Satisfied

The coding in the data set, again, is straightforward. As only one response can be selected, there is only one variable required, and the chosen value will be recorded for the variable.

Ranking questions

Earlier, we looked at questions with a range of unordered options of which respondents could select either one or multiple answers. However, there was no relative evaluation of those choices. Ranking questions permit the researcher to explicitly ask a respondent to order those responses according to a particular criterion (e.g. perceived importance or preference). Either this can be done for a full set of options, or respondents can be asked first to select those they consider relevant and then, in a second step, to rank them. Reconsider, for example, the earlier question about national identity. Imagine, an interviewee had selected 'British, Scottish and European'. They may now be asked a follow-up question:

Please consider the following characteristics, which you said describe the way you think of yourself. Please indicate which of these you think of yourself most, by ranking it number '1'. Please then indicate which of these you think of yourself second most, by ranking it number '2', and so on. Please write the respective numbers in the boxes provided.

☐ British
☐ Scottish
☐ European

In surveys conducted over the internet, the process can often be programmed to be interactive, for example, by providing respondents with drop-and-drag options to display the ranking order they prefer or by using drop-down menus attached for the options. To code this in a data set, we need a variable for each item separately, for which we could then record the respective rank.

Open format (with predetermined units)

We focussed in this section on question formats that were closed-ended, as they make up the main part of most social surveys. However, there is one set of questions where no explicit categories are offered but which are used commonly in surveys, nevertheless. These are questions where the answer can be given in a directly usable format by respondents in an open field and where we are able to use the response because the numeric answer given directly has a meaning and does not require post hoc coding. For this to be the case, it is crucial that the unit in which the response is given can be stated clearly and is easily understood by each respondent. There are a few common examples of this. For instance, we can ask respondents about their age by providing ranges of answers in a multiple-choice format (e.g. 18–24, 25–34, 35–44). Alternatively, we could also ask for the age of respondents directly as it is an easily understood concept. It is important, however, to make sure to state explicitly whether we want respondents to state their age in years or their year of birth, as only asking for age may result in people responding to the concept of 'age' in different ways. So, for example, we could ask,

Please state how old you are in years: _____

Or, alternatively, we could ask,

Please state which year you were born in: _____

The crucial consideration is that respondents clearly know what sort of answer is expected of them. Even then, a small number of respondents may enter the information incorrectly (if the data collection is not administered by an interviewer directly). So while this approach is suitable for simple concepts – like age, height or number of children under the age of 16 living in the same household – it is less suitable for more complicated concepts that could be easily operationalised differently by various respondents. The coding of the responses is straightforward. As the numeric value is informative in itself, the value itself just gets recorded.

What makes good questions – and some common pitfalls

Designing good questions requires practice and a lot of careful consideration. The expert interviewees for this volume agree on several key areas of concern that need to be addressed.

There are quite a lot of things we should consider when formulating questions for surveys, as many pitfalls exist that could impede our ability to develop survey items that can be used confidently. In this section, we will review the issues raised by our expert interviewees and which represent the most common issues that could negatively affect our questionnaire design, and we will discuss how we can avoid these problems.

Box 4.1: Ask an Expert

Good survey questions and common pitfalls

What, fundamentally, makes a good survey question?

Rachel Ormston:	You need to be able to understand it and know how you are supposed to answer it. You're looking at whether they understand what they're being asked, can they retrieve the information that they need to be able to answer it and then can they form a judgement about how to answer it? . . . It needs to be clear, it needs to be answerable and it needs to be something that people are willing to answer.
Christian Welzel:	It is very important to phrase survey items in a language that resonates with people. It has to be intuitive, because then you get a genuine, authentic answer.
Susan Reid:	It needs to be understood the same way by everyone who is answering it. So it needs to be entirely unambiguous. It's got to be very clear what it is you're asking your respondent to think about and to tell you about.
Paul Bradshaw:	[Respondents] need to be able to answer it. So the options that you give them for responding need to match what they expect to see in response to that question.

What are the most common mistakes made in formulating survey questions?

Susan Reid:	Often, answer scales don't fit the question. They're not evenly divided. So if you've got 5 points on your scale you want to have the same gap between 1 and 2 and 4 and 5. And people, especially on attitudinal scales, quite often use things like 'quite' or 'fairly' [interchangeably], and you see both of those as if they're the same thing. So it's not a scale. And questions which aren't balanced. So if you ask someone 'how likely' it is that something will happen rather than 'how likely or unlikely', you're leading people to think you're assuming it's likely.
Paul Bradshaw:	I think leading questions are a fairly common error, when you present a statement that is clearly presenting a particular position and then ask people what they think of that position but there's a value judgement in the statement.

(Continued)

Rachel Ormston: A pitfall which is really common is asking your research question directly rather than working out what questions you need to ask your respondent to answer your research question. I think the worst examples of really convoluted questions are where people have literally just taken the research question and then asked it rather than thinking, 'That needs several different questions to actually understand that'. . . . If you're doing a survey and you ask people, for example, 'What do you understand by national identity? Choose from this list', you know you've already artificially completely narrowed it down. They might never have even thought about national identity before. You need to think about what are the dimensions of national identity that you're actually interested in exploring and break that down. . . . It's not reasonable to expect members of the public to do the analysis for you, which is effectively what you're doing if you just ask them the research question. [As an example, regarding the question 'What's most important in your vote choice?'], what you're getting is their top-of-mind response to what they think has informed their voting decision. But actually quite often there are unconscious things going on in people's decision-making that they won't necessarily be able to just articulate off the top of their head. Whereas if you ask them a more detailed set of questions about how important they think different issues are and things about their personal situation, then you do a more sophisticated analysis where you look at which of these factors are actually related to how they vote, [and] you'll pick up on some of the stuff that people aren't able to articulate.

Non-discrete questions

The first criterion for a good question may sound very simple, or nearly trivial even, but is fundamental and often not easy to adhere to straightforwardly, especially when asking about complex issues. We need to make sure that respondents actually answer the question that we want them to answer. In order for them to be able to do this, the question has to be clear and focussed on one specific issue. However, it is not uncommon, especially colloquially, to conflate multiple issues into single questions. That is no problem in a conversation, where we have the opportunity to follow up on the part of the question that got ignored or where the respondent can state what part of the question they really will be referring to in their subsequent statement. But when answering closed-ended questions in particular, it would be impossible to take account of this with predefined answer options. Consider the following, very problematic example:

To what extent do you think that the reduction in trade barriers has led to economic growth and could therefore support the development of Southern welfare states?

A lot is going on in this question – too much to formulate meaningful answer options. The 'to what extent' part lends itself to using a Likert scale that could run from strong agreement to strong disagreement with a statement. But what would that statement have to be like? In this case, it would have to include the entirety of the statement, as it includes several explicitly implied causal links. So the options could look like these:

1. Strongly agree that a reduction in trade barriers has led to economic growth that has resulted in support for the development of Southern welfare states.
2. Somewhat agree that a reduction in trade barriers . . . .
3. Neither agree nor disagree that a reduction in trade barriers . . .
4. Somehat disagree that a reducation in trade barrieers . . .
5. Strongly disagree that a reduction in trade barriers has led to economic growth that has resulted in support for the development of Southern welfare states.

In other words, a respondent would have to make an evaluation about the entirety of the whole argument that is being presented, which is rather difficult. Besides the high cognitive demand and the vagueness of some of the concepts (e.g. 'development of Southern welfare states'), which permits for a variety of different interpretations, the most fundamental problem of the question is that it does not permit the respondent to evaluate the different elements of the statements separately, although they may agree with some of them but not all of them. For example,

- they may agree or disagree with the evaluation that there has been a reduction in trade barriers in the first place;
- even if they agree that such a reduction has taken place, they may now agree or disagree with the statement that this has caused an increase in economic growth (if they acknowledge that there has been economic growth in the first place); and
- finally, even if they agree with all of these assertions, they may agree or disagree that this economic growth had a particular effect on Southern welfare states.

So there is a myriad of reasons why people may disagree with this statement, but we would not be able to read off from the responses what the motivation for each person was. The question is, therefore, not a good instrument as the results from it would be muddled and of limited meaning. Partially depending on the focus of our research, there are several things we could do to improve it:

- Unless, we genuinely want to ask about the whole causal argument chain, we should break down this question into multiple sub-questions. Otherwise, if we wanted to present a holistic argument, two options would exist to properly operationalise it. We could begin by formulating a preamble along the lines of 'Some people have suggested that the following process has taken place, while others disagree with that suggestion' and could then present a clearer description of the process before asking respondents for their evaluation. Alternatively, if we are less interested in the specific case discussed here but hold a belief in the general principle, the question could be presented in a more abstract manner along the lines of 'Some people believe that a reduction in trade barriers will always lead to economic growth that results in better funding for welfare states, while others doubt that this is the case.' And then we could ask people about their agreement or disagreement with the full argument.

- In most cases, however, we would sub-divide the questions to deal with each of them separately. We would begin by examining people's views on the reduction in trade barriers, followed by a separate question on whether such a reduction had resulted in economic growth and finally a question on whether people think that economic growth is linked to welfare state issues, in a clearly defined way.
- Obviously, sometimes we may wish to present people with a particular stimulus or piece of factual information. This should be done clearly in a preamble. For example, we might be able to state something along the lines of 'Tariffs that companies and individuals have to pay for goods imported from outside the country have reduced by 15% over the last two years according to the National Statistics Office'. If we want to gauge properly whether people share the factual assessment, we would have to ask about it separately as well. Fundamentally though, we should never assume that the assumptions we might make in a statement are definitely shared subjectively by our respondents.

In sum, it is crucial that we only ask about one discrete aspect at a time when formulating a question. That could be the view on one issue or the view on the relationship between two issues, but we need to make sure we ask the question in such a way that the respondent knows clearly what we are inquiring about.

Providing non-discrete answer option sets

It is not only the question itself that needs to be clear and distinctive. The same applies to the answer options provided. Crucially, the answer options need to be discrete, which means they should not overlap with each other. This means that a particular response a survey participant may wish to give cannot be captured by more than one possible option. In simple answer sets, such as age groupings, for example, mistakes like this can easily be spotted:

Which of these ranges includes how old you are in years?

1. 18–34
2. 35–54
3. 45–74
4. 75 or above

While options 1 and 2 are neatly distinct from each other, options 2 and 3 overlap. Somebody who is 50 years old would not know whether to select option 2 or 3. The mistake would, of course, be easily rectified by changing option 3 to '55–74'. However, the problem of overlap can be more difficult to spot when the answer options reflect slightly more complex concepts. Consider the following example:

What is the highest university qualification, if any at all, you have obtained so far?

1. Doctoral degree
2. Postgraduate qualification
3. PG certificate, PG diploma
4. Undergraduate degree
5. Qualification below undergraduate degree
6. No university qualification

Several issues arise with this particular question. The most problematic one in terms of presenting non-discrete answer options are numbers 2 and 3, respectively. It appears that option 2 refers to a master's-level degree, while option 3 refers to specific postgraduate certificates or diploma qualifications but not with a master's-level degree status. The formulations do not make this clear, however. Option 2 is particularly problematic as it is very unspecific and respondents who would have achieved a qualification covered by option 3 may select option 2 as well as it covers those qualifications in the way it is formulated. Creating clearly delineable categories is important but tricky, in particular when those categories combine multiple further options (e.g. different types of qualifications here) as a group. The answer options overall are not formulated well here and are rather unspecific. It is not fully clear what is included in option 5 precisely and what is not. Two main options exist to address these problems here:

1. One could reformulate each of the categories and give them much more precise labels that make it clear to every respondent what qualifications would fall into which group. Commonly, examples would be provided (e.g. option 2 could become 'Postgraduate degree, such as MSc, MPhil, MRes'). However, even in that case, problems would remain. For example, if MA was presented as an example, it would be confusing in relation to educational qualifications in Scotland, where MA Honours degrees are given for many undergraduate qualifications.
2. Another approach would be to provide a much longer and comprehensive list of all possible qualifications that could apply and group those post hoc in the data set for subsequent analyses. If people only had to select their own distinctive qualification, no errors in grouping associations would be expected from their side. Such an approach would, of course, take up more visual space or lead to the reading out of a very long list over the telephone, so this may have drawbacks as well. Similar to the other strategy presented, however, it would substantially reduce the issue of non-discreteness the original question presented.

Providing uncomprehensive answer option sets

Apart from answer options being distinct from one another and discretely separated, the second important condition is that they should be comprehensive. We briefly alluded to this earlier, but it is worth engaging with this point in some more detail. By comprehensiveness, we mean that the options provided cover all plausible options that respondents may wish to select within the framework of the question. This does not mean that the options may be grouped or formulated precisely as each respondent would have done after reading or hearing the question, but fundamentally, they need to be able to identify one of the options provided as representing or containing their response to the question. Consider the following uncomprehensive set of answer options in relation to a question about relative national identity evaluations:

> Which of the following best describes you on a scale from 'Scottish, not British' to 'British, not Scottish'?
>
> 1. Scottish, not British
> 2. More Scottish than British
> 3. More British than Scottish
> 4. British, not Scottish

Two immediate flaws are apparent regarding the comprehensiveness of this question. There is no option allowing people to say that they are equally Scottish and British. However, this is actually one of the most common or even the most common chosen option in attitudes surveys in Scotland (ScotCen, 2018a). Reporting results without a middle option would therefore imply that a major option representing the actual views of the population was ignored. Even after including it, a second problem remains. What about people who feel neither Scottish nor British? They have no option to choose at all in this particular case. For this reason, usually a 'None of these' option and additionally an 'Other' option are added. The complete set, as it is commonly used, is an operationalisation of the commonly used Moreno question and looks more like this:

> Which of the following best describes you on a scale from 'Scottish, not British' to 'British, not Scottish'?
>
> 1. Scottish, not British
> 2. More Scottish than British
> 3. Equally Scottish and British
> 4. More British than Scottish
> 5. British, not Scottish
> 6. None of the above
> 7. Other (Write in)

So comprehensiveness addresses two important points: (1) the inclusion of plausible categories as well as (2) the possibility for respondents to explicitly state that none of the options really reflect their view. When we do not provide such options, we need to be very careful about our interpretation of the results. Closed-ended questions always ask respondents to match their thoughts to a set of predefined options. However, if we are not confident that we have presented a comprehensive set, our communication of results needs to be adjusted. Imagine that in the second case, the full question with seven options, 25% of respondents said they were more Scottish than British. Presuming we had a good-quality sample we deemed representative of the Scottish population, we would be happy to state that about one quarter of people in Scotland felt that they were more Scottish than British. We should be hesitant to make a similar statement in the first case with the incomplete four-options set. The figure for each category would likely be inflated because those who would have otherwise chosen the middle or 'None' or 'Other' category would have shifted their views to other options. Of course, we might be interested precisely in what people do when forced to make such a choice, but we should report findings in that regard then and not claim that they are a comprehensive assessment of the population.

Because of these considerations, most social surveys that aim to make statements about the levels of certain views in the population contain explicit options to appropriately capture respondents who do not feel that the answers provided suit them well for all questions. Sometimes those options are differentiated into multiple subcategories, such as the following:

- 'Don't know' (captures people who find it difficult to form a view on the question or do not understand the question)
- 'Prefer not to answer' (commonly used as an option for sensitive questions)
- 'Other (Write in)' (permits additional categories to be considered)
- 'None of the above' (explicitly enables people to state that they understand the question but do not agree with the options provided)

Recording such responses can be important beyond merely ensuring that a comprehensive set of options is provided. These responses can be informative in themselves. Knowing that people refuse to answer a question, for example, can be an indicator that the topic of the question is associated with a social taboo. High numbers of 'Don't know' evaluations might suggest that the concept presented is not clear to a wide range of people. We can even analyse these responses to examine whether certain groups of the population (e.g. those with certain levels of education or who live in certain areas) are more or less likely to not find the main answer options provided sufficient.

Leading questions

In a social survey, we want to gather people's actual views obviously. However, the way we formulate our questions can substantially influence how people respond. One of the biggest problems that we want to avoid is to be leading in a particular way. This means that we want to avoid creating a situation in which it may appear that some options are generally better options or correspond directly more closely to the question than others. Leading questions are sometimes actively used by campaigns to get people to answer questions in a particular way that is desirable for those organising the campaign. However, for proper research, we want to ensure that we avoid steering respondents one way or another. Additionally, being attentive to this in our own surveys can help us to identify cases where other surveys inappropriately provide leading questions. The following example illustrates how a question could be strongly leading:

> Do you agree that capital punishment should be reintroduced as it would help reduce crime rates?
>
> 1. Yes, capital punishment should be reintroduced.
> 2. No, capital punishment should not be reintroduced.
> 3. Don't know

The question is leading in multiple ways. First, by simply asking 'Do you agree', providing an affirmative answer is made easier compared with disagreeing with the proposition. The way the question is phrased asks the respondent to either go along with the particular idea that is being introduced or actively reject it, the latter being substantially more difficult. A more balanced way of asking such a question would be to ask more neutrally, 'Do you agree or disagree with the following statement?' The second leading bias is that the introduction of capital punishment is directly linked to a particular positive outcome, the reduction in crime rates. This frames the issue in a very particular light that may result in the respondent considering it specifically under this aspect and ignoring other concerns that may have come to mind without that prompt. This concern intersects with the issue of formulating a discrete question. There may, of course, be reasons other than a potential crime rate effect that one may have to support or oppose the idea. Importantly though, the way the question is phrased suggests that there is an empirical link between crime rates and the existence of capital punishment. We may be interested in finding out whether people actually think there is one, but we would have to ask a separate question examining that particular issue and one clearer question about whether the respondent would like to see capital punishment reintroduced (without a leading formulation).

Social desirability

When we talk with others in everyday situations, whether it is at work or in a social setting, we take into account what others may think of the things we are about to say and we may adjust our tone or even the message we plan to communicate, at least to some extent. We may even decide to censor ourselves and not say what comes to our mind because we might be worried that the issue is too sensitive or there would be disapproval of our opinion. When we adjust our response to what we think is expected from us or what is deemed as acceptable more broadly, we are trying to formulate statements that are socially desirable. Concerns about **social desirability** are not limited to everyday interactions though. Even in social surveys, we find that people sometimes do not reveal their actual views, or at least moderate more extreme positions, because they are concerned that their views might be deviant or socially questioned, even when the survey guarantees anonymity. Not surprisingly, social desirability can be more pronounced when interviews are interviewer administered, as there is direct interaction with another person. That is why, in face-to-face settings, some parts of a questionnaire, usually those dealing with sensitive issues, may be self-administered by the respondent – commonly using a computer-assisted system – to reduce the impact of social desirability (Tourangeau & Smith, 1996).

Social desirability is highly context dependent and can affect a wide range of issues. Glynn et al. (2011), for example, found that people tended to significantly underreport sexual activity in a sexual behaviour survey in Malawi, especially in relation to non-marital partnerships. The under-reporting was not distributed uniformly across the population though but was stratified by gender and age. In a study examining the consistency of self-reported sexual behaviour with actual behaviour in the USA, however, while some differences were identified, they were much less pronounced than was originally expected and were not affected by the data collection mode (Hamilton & Morris, 2010). This suggests that we need to be very careful about making assumptions regarding social desirability patterns in different contexts. We should not simply assume that a pattern observed for a particular population may also work in the same way for a different one.

Responding in a way that addresses social norms is not just limited to sensitive issues, such as sexual behaviour. Studies have shown that more people report having voted in an election than the actual turnout rate would suggest (Belli et al., 1999). This is a problem for researchers who try to use sample-based surveys, for example, in election studies, to make inferences about the political behaviour in countries. Understanding the mechanisms of such biases is therefore very important. While over-reporting of electoral participation is common, for example, the extent is context dependent. Karp and Brockington (2005) demonstrate that the level of over-reporting on this issue is dependent on the actual participation rate. In countries where voter turnout was higher, they found that the over-reporting effect was greater, suggesting that when one's own behaviour is more deviant from the overall norm, the pressure to respond in a socially desirable way could be greater. Being aware of social desirability issues is important not only in the questionnaire design phase but also later at the data analysis stage, where comparisons to official data, where available, can be a meaningful way to assess the extent to which results may have been affected.

Thinking about the questionnaire structurally

So far we looked at the design of individual questions and discussed a wide range of issues that could affect their adequacy and usefulness. However, when designing the questionnaire for a survey, we need to consider structural questions in addition to ensuring that each question is formulated well in its own right. Survey participants do not respond to questions in isolation but in the context of which questions they have engaged with before already. Furthermore, for our analysis, sometimes the results from an individual question per se may be less important to address our research questions than the interplay between the results from multiple items. We therefore need to make sure that all the elements we require for the analysis are actually contained appropriately in our questionnaire.

The effect of question and answer option order

Thinking carefully about the order in which survey participants are asked to answer questions is very important, as it may affect the results (Strack, 1992). When answering particular questions, people can, but do not have to (McFarland, 1981), be affected by the previous ones they have considered, as the topics brought up may affect the associations that are established when confronted with subsequent questions. **Question order effects** can occur for a wide range of survey themes and across different modes of survey administration. Lasorsa (2003), for example, finds that people reported lower levels of political interest if they were first asked a set of challenging political knowledge questions that may have made them feel less positive about engaging with political issues.

Ordering effects do not only apply to the sequence of questions but also matter for the response options provided. The position of answers within a set of answer options may affect the likelihood of those responses being chosen and therefore the subsequent relationships between variables (Krosnick & Alwin, 1987). Let us reconsider one of the example questions from the beginning of this chapter asking respondents about their religious affiliation:

Which of the following religious affiliations would you say you belong to, if any at all?

1. I do not have any religious affiliation.
2. Catholic
3. Protestant
4. Islam
5. Judaism
6. Other (Please write in your affiliation)

The question effectively asks about two things: (1) whether respondents have any religious affiliation and (2) if they do, what that affiliation is. Ideally, we would separate this out into two separate questions, where only those who say they have an affiliation would be asked about which one applies. The concept is fairly straightforward, so therefore combining the options here can be feasible because space on the survey for additional questions is limited. In such a case, however, it is common to then place the option that is the most different, in this case not having any affiliation at all, at the beginning of a set of possible options, to signal to the respondent that it is easily available and can be chosen. Another strategy often employed to address answer option order effects is randomisation of the order. For example, when asking about preferences for political parties, the order in which respondents get to see the lists could be randomised throughout to compensate for list order effects. Similarly, when presenting lists with multiple options or ranking exercises, order randomisation of answer items can help ensure that no biases arise in favour of items higher up on the list. While not all questions or answer option sets may be affected equally by ordering effects, considering them during the design phase of a questionnaire is of great importance to avoid potential biases.

Box 4.2: Case Study

Asking about the independence of Scotland

Question order effects are sometimes actively used to create responses that are conducive to a particular marketing or political message, but they distort the actual views people hold. A prominent example is a poll that was conducted by the Scottish National Party (SNP) in the lead-up to the 2014 referendum on Scottish independence. In September 2013, they published the results from the poll, conducted by the polling company Panelbase, which reported that the independence movement was ahead in a poll for the first time, with 44% supporting Scotland becoming independent, 43% opposing it and 13% declaring that they were undecided (Panelbase, 2013). Obviously, the 'Yes' campaign was very happy with these results, as they could claim that they had the upper hand and the momentum was shifting in

(Continued)

their direction. Newspapers ran the story with headlines such as 'SNP Poll Puts Yes Campaign Ahead' (Gardham, 2013). The results were particularly impressive as no poll published so far had recorded a plurality of people saying they would vote for independence.

Support in other polls conducted in August 2013 ranged from 25% to 37% only (ScotCen, 2018b). So the findings from this survey should appear to be somewhat peculiar. The reason why support apparently 'increased' so much, but was not replicated in any other surveys conducted during the rest of 2013 (which were not funded by the SNP), can be explained when we look at what people were asked before they were requested to answer the offi-cially approved question 'Should Scotland be an independent country?' The first question of the poll was 'Do you agree or disagree with the following statement: "Scotland could be a successful, independent country"?' – which 52% agreed with while 37% disagreed. So even some people who did not favour independence responded affirmatively to this question. The question clearly creates an image that frames independence in a positive light. This was then followed by the question 'Who do you trust to take the best decisions for Scotland: the Scottish government or the Westminster government?' – 60% chose the Scottish option, and only 16% selected the Westminster option. Taking both questions together, the survey par-ticipants were primed to think positively about decision-making being located in Scotland. Asking about their views on independence after those two questions had the potential to influence respondents and might explain why this particular poll was such a pronounced outlier compared to all other polls conducted in the same period.

Linking questions for analyses

We design questionnaires so that we can undertake particular analyses that enable us to answer the research questions we set. These questions often imply a degree of complexity that cannot be captured by single-item questions. One question usually can only reveal a very specific attitude or evaluation, but research questions or the concepts they try to work with require a greater degree of depth. For example, we may not simply be interested in people's views on whether a particular economic policy should require greater or lesser state involvement (in comparison to private sector involvement), but we may be interested in more broadly assessing how people evaluate the ideal degree of balance between state and individual responsibility. We could imagine questions that try to ask about this at a high degree of abstraction, but those questions would not be likely to actually reflect how people think about these issues. Instead, we might need to ask a range of questions that cover different aspects of this overall concept in the first place, ensuring that respondents are giving answers to questions that are more relatable to their thoughts and experiences. Such batteries of questions can often be helpful to engage with more complex issues that we cannot neatly ask about in one single question but that we can gain an insight into by asking for a greater number of more discrete evaluations. For the issue raised above, consider the following example (Table 4.2) of questions from the World Values Survey (2012):

Now I'd like you to tell me your views on various issues. How would you place your views on this scale? 1 means you agree completely with the statement on the left; 10 means you agree completely with the statement on the right; and if your views fall somewhere in between, you can choose any number in between.

Table 4.2 Example questions from the World Values Survey (2012)

1	2	3	4	5	6	7	8	9	10
Incomes should be made more equal					We need larger income differences as incentives for individual effort				
Private ownership of business and industry should be increased					Government ownership of business and industry should be increased				
Government should take more responsibility to ensure that everyone is provided for					People should take on more responsibility to provide for themselves				
Competition is good. It stimulates people to work hard and develop new ideas					Competition is harmful. It brings out the worst in people				
In the long run, hard work usually brings a better life					Hard work doesn't generally bring success – it's more a matter of luck and connections				
People can only get rich at the expense of others					Wealth can grow so there's enough for everyone				

All of these questions deal with different aspects of views on the functioning of markets and market principles in the relation between people, business and the state. Trying to summarise all of these in one meaningful question that could capture all domains would appear to be impossible. Additionally, this breakdown permits us to actually conduct analyses into the concepts we are interested in. While we have been assuming so far that all of these issues would be related, using a set of questions like the one in Table 4.2, we can analyse whether that is the case and, if they are, how precisely they relate to each other. For example, we may hypothesise that people who say that they favour more government involvement evaluate market principles, such as competition, more sceptically and do not think personal profit motives achieve much good. In that case, we would expect that all six items above would be strongly correlated to each other, so knowing what a person says for one issue would be a strong indication of what they might say for any of the other issues. We would theorise that there might be one underlying latent concept for all of these variables: a belief or scepticism in the functioning of free-market principles. However, other hypotheses are also plausible. While we might expect that the items that explicitly ask about the role of governments would be related, we may not be convinced that those perceptions are perfectly related to evaluations of how individuals behave in markets. For example, somebody may think government involvement is positive and evaluates those items similarly but may not therefore be sceptical about individual behaviour in a marketplace or think that competitive behaviour has to be negative. In that case, we might theorise that there are two separate latent constructs: one underpinned by correlated items about government, reflecting people's views on the state–market relationship, and the other underpinned by items about the economic behaviour of firms and people, reflecting respondents' views on the economic patterns between non-state actors. Undertaking so-called dimension reduction (e.g. using exploratory factor or principal component techniques) or latent variable analyses (e.g. confirmatory factor analyses) allows us to empirically answer these questions when we have such question batteries. The fundamental idea behind such approaches is to examine the structures that may connect the responses of a number of different questions. *Exploratory* approaches tend to look at the correlations between all the question responses we are interested in to identify patterns of responses that may suggest which questions could be grouped together. *Confirmatory* approaches start by hypothesising particular relationships, stipulating certain underlying, latent variables that may be reflected by the answers to a certain set of questions and examining whether those relationships do indeed exist. A detailed engagement with these analytical techniques is beyond the scope of this volume, however.

Validity and reliability concerns

Fundamentally, we aim to measure correctly the things we aim to examine. What we try to do to achieve this is to ensure both *validity* and *reliability*. *Validity* refers to the question whether we are measuring what we actually claim to measure, while reliability refers to the consistency of the results we obtain. When we design a survey, we ultimately take a particular concept and try to operationalise it in such a way that we are able to measure people's responses that reflect their evaluations in relation to that particular concept. In our example above, we are interested in studying how people position themselves in terms of whether the government or individuals should take responsibility for what is going on in the economy and society. We could simply ask that question directly, but the concept is rather abstract, and if people just responded to that question, they may have very different views about what the question actually means. We might not get responses that actually correspond well to the concept we want to examine, so there would be a mismatch between our conceptualisation and our operationalisation.

Instead, the complex concept of government versus individual responsibility was broken down into several smaller sub-questions that people are able to answer more straightforwardly. Rather than having to make a statement overall about the big abstract questions, people are asked about more direct things, such as whether they prefer more or less inequality or more or less public versus private ownership. Taken together, the answers to all these questions will give us a good indication of where people might stand in their views of government versus personal responsibility. This also allows for nuance, if people might prefer personal responsibility in some domains but government action in others. Overall, the responses would have a greater degree of validity in relation to our concept of interest compared to us just asking one highly abstract question. We can further distinguish between *internal* and *external validity*. Internal validity focusses on an assessment of the survey instruments themselves. Are they constructed in a way that addresses the concerns we discussed in this chapter, so that we can meaningfully claim that they provide a good measure of our concepts of interest? This includes many domains and in our discussion above would, for example, examine whether the construction of government versus personal responsibility preferences was balanced and not leading one way or another. *External validity* requires us to examine whether what we measure actually corresponds to the concept of interest. For example, we may have other questions in our survey about evaluations of the economic system. If our survey instruments were developed well, we would expect that somebody whom we categorised as being focussed on individual responsibility based on their responses to the questions above would provide responses to questions on other economic concerns that represented an economically more right-wing view as well.

Reliability is concerned with the question of whether the results we obtain are consistent, in relation to both the responses from a survey participant (*internal reliability*) and the study overall (*external reliability*). When we ask questions of a respondent, we expect that the answers are meaningful and reflect their actual attitudes. So therefore, we would expect that if we asked the same question again at a later stage, the responses would be the same (or at least very similar) – unless we designed questions that are intentionally sensitive to changes in the environment (e.g. the time of the day) to capture information about certain external influences. So if we administered the same question multiple times and the respondent gave us very similar responses, we would conclude that there was a good degree of *internal reliability*. We can think about reliability at a larger scale as well. The idea behind *external reliability* is that if our study was replicated in the same way (e.g. by another researcher) – for example, the same survey questions were asked of a different representative sample of the same population – we would hope that the results of the study would be similar. If they were not, then the reliability of the study overall could be questioned.

Taken together, these considerations matter when we decide on how to construct our questions and in particular whether we can ask single questions or might want to consider combining the responses from multiple questions. Even at a simpler level, sometimes it can be better to not ask a question directly on the issue we want to examine but instead to indirectly assess a topic, even if that may seem counter-intuitive at first. A key example of this are questions trying to determine what topical issues are decisive in determining what party people are voting for in an election. The most straightforward way would be to simply ask people what issue is most important to them in making their electoral decision, which is commonly done in polls. However, that approach, while seemingly plausible, is flawed. In answering that question, people may not refer to what actually drives their vote choice (as it is quite hard to untangle the many influences on voter decisions cognitively) but instead may refer to what issues they consider most salient because, for example, those issues had been at the forefront of media debates or campaign efforts. In other words, the question reveals what people *think* affects their vote choice. It is less suitable to assess what actually *correlates* with vote choice. In order to assess this, we actually need to separate the party choice from the evaluation of a particular issue. So we need to ask, on the one hand, what political party people would choose and, on the other, what their views are on a wide range of issues. We can then look at the correlation between the two. If those who support different parties have very similar views on a political issue, we know that this is not actually a decisive issue as it does not differentiate party supporters. The issue we find to show the greatest divergence in views between different party supporters is the one that indeed appears to be the most decisive (and one that campaigns may wish to focus on in developing their messaging).

Conclusion

So in designing a questionnaire, not only do we have to think about how to formulate appropriate questions in their own right, but additionally, we need to consider how we plan to analyse those questions and whether those analyses would be sufficient to address our overarching research aims. Sometimes that means understanding that we cannot engage with an issue through one discrete question but instead need to design multiple questions that may even be situated in different parts of the survey.

Chapter Summary

- Designing questionnaires is complex and hard work. It usually involves many iterations of discussions about ideas for questions and possible formulations (the process will be discussed more in Chapter 5).
- Fundamentally, we need to make sure that our questions allow us to actually measure the things that we intend to examine. In order to achieve this, they do not only need to correspond well to the concepts related to our research questions, but they also need to be understood clearly in the same way by the respondents who take part in the survey.

- In addition to the formulation of the question, the answer options also have to make sense, and we need to carefully keep in mind what the particular set of options we have provided suggests for our interpretation of the results.
- Ultimately, we always measure what people say in response to the particular question framework we present them with. It should never be assumed that the responses are at the forefront of people's minds per se. They may not have thought about the topic we ask them about at all that day, so any response should always be understood within the specific context of the survey we administer.
- This also means that we should not only think about questions in isolation but also consider how the order and grouping of questions in a survey may create framing influences that could affect how people respond to later questions.
- To ensure that our design works in the way we intended, thus allowing us to achieve high validity and reliability, it is usually a good idea to perform some tests before actually launching the survey in the field. Chapter 5 will discuss how best we can do that.

Further Reading

Beatty, P., Collins, D., Kaye, L., Padilla, J., Willis, G., & Wilmot, A. (Eds.). (2019). *Advances in questionnaire design, development, evaluation and testing.* Wiley.

This edited volume discusses the current state of questionnaire design methods and goes beyond the basics. It provides detailed insights into different ways of testing and validating questionnaires and is useful for anyone wishing to deepen their knowledge on that part of survey design.

Jann, B., & Hinz, T. (2016). Research question and design for survey research. In C. Wolf, D. Joye, T. Smith, & Y. Fu (Eds.), *The SAGE handbook of survey methodology* (pp. 105–121). Sage.

This is a very good text that engages with the crucial issues around designing the questions used in surveys in a concise manner.

5

ENSURING SURVEY QUALITY: PILOTING, CHECKS AND CROSS-CULTURAL COMPARABILITY

Chapter Overview

Introduction 82
The complex path to a good-quality survey 82
Pre-fieldwork quality checks 87
Post-fieldwork quality checks 94
Further Reading 100

Introduction

As we have seen in previous chapters, developing a survey well takes a lot of time, thinking and resources. There are many points that need to be considered and many things that can go wrong if researchers do not address the concerns in sampling and questionnaire design outlined in this volume. However, we never know what the composition of our sample ultimately will look like and how the respondents will answer the questions we pose until we get the final data set – which is often a moment filled with excitement and nervousness in equal measure for the researchers involved. Before going into the field, researchers therefore normally aim to test as much as possible how well their survey is going to work in line with their goals and expectations. Several things can be done in order to spot problems and mistakes before one actually begins the fieldwork. In this chapter, we will review some of those approaches that are frequently undertaken by survey researchers following the initial development of the questionnaire and sampling strategy. But even after the data from fieldwork arrives, researchers do not simply begin to work with the data and conduct the analyses. Instead, usually a wide range of checks is undertaken to examine the data and make sure that it meets the quality standards one aspires to. This potentially leads to adjustments that need to be made to the data or the analyses to achieve an adequate quality for analysis, so we will also briefly look into some of those issues (although they are discussed in detail in Volume 5 of this series, *which focusses on secondary data analysis*). Furthermore, surveys are often conducted in order to compare findings across different regions or even countries. When that is done, there are particular concerns in relation to cross-context comparability (e.g. the validity of concepts in different countries or issues of translation). We will highlight some of those core concerns in this chapter.

The complex path to a good-quality survey

Each consideration in relation to survey design requires complex judgements. Bringing all of it together in a meaningful way, however, is a substantially larger challenge. While quick polls using established questions and applied to a preselected panel of respondents can be undertaken over a few weeks or even days, designing and administering a comprehensive survey can take months or even years when it is conducted in a wide range of countries. Many steps have to be taken sequentially but often also in parallel. The case study below summarises the details for a high-quality political attitudes survey, the Scottish Social Attitudes Survey. Using it as an example allows us to get an overview of some common issues that survey developers have to deal with in order to achieve high quality.

Box 5.1: Case Study

The processes required to create the Scottish Social Attitudes Survey

(Based on an interview with Susan Reid and Paul Bradshaw)

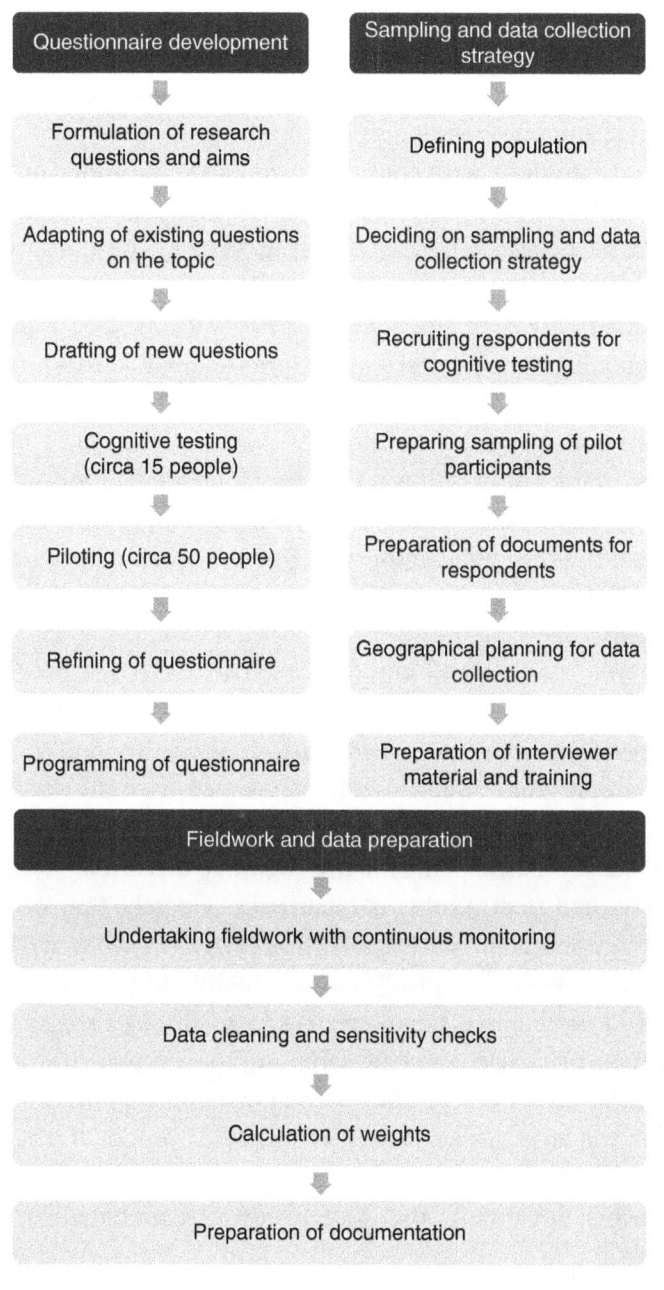

Susan Reid and Paul Bradshaw from ScotCen Social Research, who are responsible for the survey, outline two main streams of activities that often occur in parallel and intersect with each other. The development of the questionnaire is, of course, a crucial task. As we discussed in previous chapters, it is crucial to have a clear idea of the goals that are to be achieved by the survey and the questions that need to be answered. Especially when surveys are developed for or with partners or funders, it is very important to ensure that the aims are clearly conceptualised so the survey development is adequate. Initially, researchers would then typically look to existing surveys covering similar topics, especially those that are considered strong methodologically. Questions that are tried and tested can be replicated meaningfully with confidence and have the added quality that they allow for comparisons with other surveys – for example, for sensitivity checks, as we will discuss below. It is rare, however, that all the questions one wants to address have already been asked well elsewhere; therefore, new questions will need to be drafted, discussed and edited by survey teams. It is thus very useful, if possible, to subsequently undertake some form of cognitive testing with a small number of respondents (discussed below) to explore whether the questions are actually understood by people in the way that the researchers intended. After readjustments, a larger pilot of the actual survey, using the proper administration method chosen, can be carried out to examine how it fares in the real context of fieldwork. All the insights together can then be brought together for the final programming of the questionnaire for whichever method of application, ready to go into the field.

At the same time, the work for sampling and data collection has to be prepared, so that everything can be in place for the fieldwork on time. Based on the goals identified, a population has to be defined well, which then allows for the sampling and data collection strategy to be developed with all the considerations outlined in Chapters 2 and 3. However, before any respondents can be recruited for the actual survey, participants for the cognitive interviews need to be identified and approached first. As this is qualitative research, they are not required to represent the population statistically; however, the goal is always to provide as wide a range of individual profiles as possible within the group that participates in order to get many different people to go through the survey questions and provide their feedback. Also, the sampling for the pilot has to be prepared; depending on the nature of the pilot, this may involve going through the exact mechanisms of the later survey or slightly simplified processes (the differences of which we will discuss below). While one team works on the questionnaire, there are multiple other documents that also need to be prepared for respondents, including information and consent sheets. If respondents are invited to take part in advance, this may also include advance letters or other communication for respondents (e.g. in mixed-mode surveys). Before individuals can be approved, the detailed geographies in which the surveys are conducted have to be established by fieldwork teams. In the case of face-to-face administered surveys, like the Scottish Social Attitudes Survey, this involves the selection of particular areas and specific starting points, for example. Finally, the interviewers conducting the work have to be trained and the materials for them prepared as well (which we will discuss in more detail below).

Only once all of those steps are completed and brought together can the **fieldwork** actually begin. During the fieldwork process, teams will constantly monitor incoming data, the distribution of respondent profiles and also the differences between interviewers or regions in terms of response rates and patterns. This helps identify potential problems in the practical implementation that can be adjusted before larger problems affect the data collection over longer periods. In quota-sampling approaches, this monitoring is essential as well, as it is the main way for researchers to affect the distribution of respondent profiles. Once the data is received, survey specialists need to undertake a range of key checks to 'clean' the data in the first instance (e.g. by removing cases where errors have occurred or identifying data entry mistakes). Furthermore, they will conduct in-depth sensitivity checks to appraise the quality of the sample and assess errors that have occurred in the sampling. Accordingly, weights will be calculated to account for divergences from expected distributions. The last step before the data is ready for analysis or publication is the writing of a detailed documentation, which each good survey should have and which outlines the details of the processes discussed here, including an account of sample compositions and the approach to weighting, so that any future user of the data can examine the data properly.

As the above summary clearly shows, conducting a high-quality survey, like the Scottish Social Attitudes Survey described in the case study, requires a lot of effort and complicated considerations by many involved people. However, as Reid points out, being a relatively straightforward attitudes survey means that it is not as complicated as some other ones which are conducted over repeated time points, with multiple members of a household or by collecting physical data about respondents' health. In her own words, '[The Scottish Social Attitudes Survey] is probably the simplest survey we run. There are probably a lot fewer steps than in some of our more complicated ones.' Survey development, done properly, indeed is a complicated endeavour and in the following sections, we will look at some of the key steps in more detail.

Box 5.2: Case Study

Young people's political engagement

In the context of the reduction of the voting age to 16 years in the Scottish independence referendum 2014, a unique survey was conducted to assess the political attitudes of those under 18 years of age, who had traditionally been excluded from political attitudes surveys as they were not eligible to vote. In designing the questionnaire (Eichhorn et al., 2014), the researchers used many established questions from existing high-quality surveys, such as the Scottish Social Attitudes Survey. However, to check whether the questions worked well for younger respondents, a pilot was conducted in which the draft questionnaire was administered to about 100 young people in a school, after which they were asked about their engagement with the questions to identify problems in interpretation and responses. Based on this, several questions were altered or dropped, illustrating how important piloting can be to identify potential issues. Some examples of such changes are presented here.

> Original question: If Scotland were to become independent, would you feel pleased or sorry or neither pleased nor sorry? If pleased or sorry: Is that very or quite pleased or sorry?

The question works well for adults overall, reflecting a general mood about the potential outcome of the referendum. But many young people in the pilot voiced strong objections. When asked what they found problematic, one student summarised the issue well by asking, 'Why should I be sorry for something that isn't my fault?' Clearly, several young people associated 'being sorry' with being told that they 'would be sorry' if they did something wrong, so the interpretation of the question was a very different one to that by adults. The question was therefore not used in the survey.

> Original question: Thinking about schools, do you think that
>
> 1. schools should be the same for everyone in the UK or
> 2. the Scottish Parliament should be able to decide what Scottish schools are like?

Normally, when this question is answered in surveys, people think about 'schools' at a higher level, effectively synonymous with 'school education' as a policy area; they are happy to decide on at what level decisions should be made. Young people have no problem with the concept of the question per se, but they think about schools in a very different way. In the pilot, they asked what aspect of schools specifically they were asked to comment on. That makes sense, as school is such an important aspect of their daily lives that they do not evaluate it in the abstract but in very concrete terms. So instead of talking about schools in the abstract, the final survey used a very specific aspect of schooling, which meant that all respondents would evaluate the same thing. 'Schools' was replaced with 'school education and in particular the curriculum and exams'.

> Original question: To what extent do you agree with the following statement: 'Sometimes politics and government seem so complicated that a person like me cannot really understand what is going on'.

This question works fine in standard surveys. The 'person like me' part of the question is meant to relate it to the respondent, but for the young people in the pilot, it had the opposite effect. Many said that they felt rather patronised by the question as they thought it suggested that they individually may not be able to understand things. In the final survey, 'a person like me' was therefore replaced with 'young people' to provide a more generalised formulation.

Pre-fieldwork quality checks

As discussed above, there are several crucial stages that are ideally undertaken before putting a survey in the field for data collection to ensure that the questions developed actually work. In this section, we show why such checks are important and present four key processes: (1) cognitive interviewing for new questions, (2) piloting, (3) interviewer preparation, and (4) translation, when conducting surveys in multiple countries or in more than one language.

Cognitive interviewing

When new questions are developed for a survey, it is difficult or even impossible to know how precisely a respondent confronted with the questions raised may understand, interpret and subsequently respond to the question. Being confident that the understanding we assume as a researcher is also the one shared by respondents is crucial, however, if we want to be confident in our interpretation of the results. **Cognitive interviewing** is a qualitative method that allows a questionnaire developer to gain insights into these issues. For cognitive interviewing, a small number of respondents are selected who will be meeting with an interviewer, usually in a one-to-one setting. They will be presented with a set of questions that have been developed for a particular survey and asked them, ideally in the same mode as the survey (i.e. on a screen if it is meant to be administered in that way or read out by an interviewer if it is a face-to-face mode). Instead of only recording the answers to the questions, the interviewer will then listen to the respondent as they describe their engagement with the question to ascertain what they thought about while answering it.

There are two dominant approaches that can be distinguished as to how cognitive interviews are conducted: *thinking-aloud* and *probing* techniques (Beatty & Willis, 2007). Thinking-aloud approaches refer to a method of administration where the interviewer takes on a very passive role, whereas the interviewer plays a very active role in probing, where detailed questions are asked to examine the points raised by participants in more depth. Effectively, it means that the respondent is instructed to talk through the process of answering a particular question, with some advice on what sort of things they may generally want to comment on. The interviewer's role then mainly is to ensure the correct administration of the questionnaire and to remind the respondent to continuously 'think aloud' and share their thoughts about the process, while the interviewer makes notes about the comments. In the alternative approach, the interviewer takes on a much more active role. While initially the respondent may provide their own views unfiltered, the interviewer may then ask probing questions to develop a deeper understanding into the reasons for particular choices or to enhance the understanding of particular formulations a respondent may have used that were not as precise as they could have been. Obviously, the extent of involvement of the interviewer can vary even when comparing different probing approaches, and there are mixed forms and specialist alternatives available for particular contexts too (Willis, 2005), including group methods that are discussion based (Bowden et al., 2002, p. 327).

While the technique is popular and very useful in detecting problems in questionnaires (Drennan, 2003), there are several considerations of caution that should be taken into account. Beatty and Willis (2007) point out that often very little attention is paid to the precise size and composition of the sample that is interviewed (p. 295). While researchers using cognitive interviewing typically do not claim that it is representative of the population, as surveys ultimately are meant to be, it is nevertheless important to keep in mind that this means that the technique cannot capture all possible dissonances and problems for all groups of the population. Even studies with large samples (more than 50) have shown that additional issues can be identified after many interviews, so typical numbers (10–20) are unlikely to meet a full saturation point. Cognitive interviewing should therefore be seen as most useful in identifying major issues that are likely to affect large groups of the population. Another key area for caution is the approach to probing. Willis (2005, p. 115) points out that it is crucial that we avoid probing that leads to the identification of problems that do not actually exist. This can happen when probing questions induce respondents to think about problems that actually did not affect their decision but that may seem plausible post hoc to them once introduced as a topic. Similarly, interviewers themselves may over-construct a problem based on the observations made and need to be careful in how they interpret the responses from participants. In addition to this issue of subjectivity inherent in a qualitative approach, some have also pointed out the artificiality of the process, given that it does not reflect how people would normally discuss such topics (Drennan, 2003). When this is seen as a major problem, some researchers have employed field-based cognitive interviewing techniques in addition to those used in experimental settings (DeMaio & Rothgeb, 1996). Despite these considerations, the insights gained from cognitive interviews are likely to enhance the quality of the questionnaire, providing further input into the editing process of the questions that are meant to be administered.

Piloting

Once questionnaires are developed, ideally with some form of qualitative checks for the questions, such as through cognitive interviewing, it is important to see how a survey actually works when administered in the field. It may not always be possible, for example, due to budget constraints, to undertake extensive qualitative testing of questions in advance; however, researchers should always strive to perform some form of pretesting that provides feedback on the questions before **piloting** the actual survey. If cognitive interviewing techniques aren't feasible, at least expert evaluations from people not directly involved in the development of the survey may be obtained (Bowden et al., 2002, p. 327). A proper pilot will not typically permit researchers to gain the qualitative insights discussed above, so it is ideal to triangulate feedback from different insights.

When conducting a pilot, the main goal is to investigate how the actual administration works in the field. So the sample selected for the pilot should ideally already approximate the one that will ultimately be selected, following the same methods of sampling and data collection. For large surveys, Dillmann (2000) suggests that a good pilot sample would consist of 100–200 respondents to permit meaningful analyses; however, typically, standard population surveys often generate smaller samples of closer to 50 respondents, in particular when data collection modes are more expensive, such as in face-to-face collection methods.

In practice, survey developers typically distinguish between two approaches: straightforward piloting and full *dress rehearsals* (a term used fairly commonly by practitioners; see, e.g. Moser & Kalton, 1992). While a pilot will be administered using the same methods selected for the overall project, the focus is largely on an analysis of the results. This means that once the data are available, distributions of responses are checked – in particular to identify problems of non-response, high levels of 'don't know' or 'refused' kind of responses or dropout rates, for example. In a dress rehearsal, the scope of the pilot is much wider. In those instances, all aspects of the data collection are scrutinised in detail, including the transmission of data and its storage, all aspects of the administration of the sampling technique, and all 'back-end' processes. Effectively, in that case, the team goes into a fully operational survey mode but stops after a certain sample size has been reached to then evaluate the insights and make adjustments to any part of the process.

Especially in the case of a dress rehearsal approach to piloting, insights about the process are gained not only from the data collected quantitatively but also from those who are involved in carrying out the survey. In the case of interviewer-administered surveys, that means discussions with the actual interviewers, who, in addition to carrying out the normal survey, will have been asked to make notes about the respondents' reactions to questions (Bowden et al., 2002, p. 328) – for example, when respondents ask for additional clarification (potentially indicating that a question is too complicated in its phrasing). When, on the contrary, surveys are self-administered by respondents (or parts of a survey are carried out in that way), a key feature in relation to pretesting and piloting is the usability (Willis, 2016, p. 364). For example, in online surveys, the way people interact with the particular interface is crucial. Pretesting may involve forms of cognitive testing with probing questions about the engagement on-screen; however, advanced techniques can also involve eye tracking to see how respondents react to particular stimuli and whether they properly read or skim certain sections. Checking the data after a pilot will also reveal whether any routing that has been developed (making the selection of what question is asked next depending on the answer to a prior question) works properly and respondents are being shown the questions they are meant to address.

Taking together the information from whichever form of piloting is utilised provides a unique opportunity to improve the quality of the survey and avoid any pitfalls that may otherwise lead to rendering certain parts of the survey ultimately not usable because of particular mistakes. Together with forms of cognitive interviewing, it also allows us to increase our confidence that ultimately we will be able to interpret the data we collect in the way that we anticipated the questions could be used.

Interviewer preparation and briefings

When surveys are administered by interviewers, their role is very fundamental to the quality of the data collection. Interviewers can affect respondents substantially in two crucial ways (Blom, 2016). First of all, depending on how well the interviewer establishes first contact with a person and issues the request to participate in the survey, the likelihood of the potential respondent actually taking part in the study may differ substantially. This can be of particular importance when interviewers try to recruit people face to face and, if things do not go well, when they might be more comfortable and effective, for example, in contacting people similar to them, which could result in differential participation rates for different population groups, thus distorting the sample composition. Second, while conducting the interviews, it is crucial that interviewers carry out the work in the same manner for all respondents, so that the results are comparable. That may include not providing additional information, even if respondents ask about a question, other than anything that has been pre-approved and would be usable in all interviews. This is particularly important as most surveys are carried out by multiple interviewers. Not only is it important to ensure that they are consistent in their own work, but it is also necessary that all interviewers carry out the work in the same way. Such 'interviewer effects' (Blom, 2016, p. 392) are to be minimised as much as possible. They are often something to be looked at in the sensitivity checks carried out after data collection and, ideally, are also considered during the piloting phases.

Ideally, however, interviewers are prepared in advance of carrying out their work to minimise the emergence of interviewer effects in the first place. Interviewer training is considered crucial in this regard and can range from providing clear written materials to group sessions with interviewers and survey developers to discuss the project, establish joint understandings and go through trial runs. When such training is developed using the findings from empirical research into how interviewers may affect outcomes, the efforts can indeed have positive effects on the participation of respondents (Groves & McGonagle, 2001). Trained interviewers have been found to achieve better response rates and manage to obtain more information, especially in relation to questions that require more substantial interviewer–respondent interaction (Billiet & Loosveldt, 1988).

Interviewer trainings, in particular, if not conducted remotely (e.g. through video conferencing) but in person can be rather expensive and an additional resource requirement for a survey project. While training is therefore often limited, Blom (2016, p. 386) argues that it should be considered a crucial part of any survey data collection and is basically always likely to improve the quality of the survey by reducing inter-interviewer variation. Despite its positive effects, even very rigorous interviewer training does not preclude any forms of unwanted interviewer effects (Kish, 1962), so when working with multiple interviewers, it is important to monitor their progress carefully.

Box 5.3: Ask an Expert

Susan Reid and Paul Bradshaw: On cognitive interviews and pilots

Why is cognitive interviewing important, and how does it work?

Susan Reid: I'll give you an example of something I'm working on to help explain. We're trying to ask people about their attitudes to young people who've been in care. So everyone is sitting around the table who is working in that field; they know exactly what we're talking about. However, you design a question which talks about young people in care, and you start thinking, 'How is a member of the general population going to understand that phrase?' Actually they could think about people being in care because they have a disability; so maybe they need care in that way rather than necessarily because of the situation that they've had at home. So the cognitive testing phase is important because it allows you to step outside of your 'I know what

(Continued)

I'm talking about because I'm sitting in a room with policy experts' space and into the real world and say, 'I'm going to talk to some real people to try and work out how much they know. And if they don't know very much, can they still answer the question?' So it allows you to understand in depth how the general population understand your question. . . . And the process is that you'll read the question out and they'll answer it as they will in the main survey. And then you stop and say something like 'Tell me in your own words what that question was about'. Or 'Describe to me what this question said. What did that bring to mind for you?' And then, 'When you answered "strongly agree" to that question, how did you come to that answer?' It's really about understanding people's thought processes as they go through the question, so that you can then work out if different people are understanding it differently and where you might need to give them more information.

What are the issues you look out for and examine when you undertake survey pilots?

Susan Reid: So I think we use piloting differently on different surveys. On SSA [Scottish Social Attitudes Survey], it's very much about questionnaire design. What we're looking out for in the findings are the answer distributions. When we're asking attitudinal questions, we don't want a lot of questions that everybody is answering the same way. The point of the questions is trying to get to that tipping point where you've got a range of responses. And we're also asking the interviewers to pick up where respondents seem to be struggling with the question or if they ask to repeat it. Or if they're giving inconsistent answers to questions which are trying to tap into a similar underlying attitude, then we'll realise that those questions aren't working.

Paul Bradshaw: [For SSA], they don't need a dress rehearsal because the dress rehearsal is about replicating everything that's going to happen at the main stage of the survey. So all the processes, everything from you sending out your advance letter, to knocking on doors, to receiving your work packs, to handing over a piece of paper, to transmitting your data, to the processes of doing heights and weights for people, . . . getting the consent and so on. It's ensuring that you haven't missed something that's critical for your survey to be able to operate in the way that you expect. . . . If we've designed a letter, do people get the letter? Did they know what it was talking about; do they remember receiving it? If we give them a leaflet, do they understand the leaflet?

When piloting a survey that is administered online, what can you look out for in the pilot?

Paul Bradshaw: If we're sending out an email with a link, we can check how many people open the email, click through on the link and then go on to complete the survey. We would look at the rates of response to individual questions and to a set of questions on-screen, because

> often online surveys will have a grid of questions. And it might be six items with five responses presented in a grid. We're looking, for example, at whether people have a tendency just to give the first response all the way down. Are some questions more likely to have been missed than others? Have people refused to answer it, or have they answered, 'Don't know'? And we'll look at duration, how long it takes to complete it.

Susan Reid: And we do testing with web surveys when we are in the room with them, combining it with a cognitive test but seeing how they click on it and then again asking them, 'How did you choose your answer?' And realising they hadn't seen the two other answer options because they weren't all underneath each other, so understanding more about the layout and its influence.

Translation

When a survey is conducted in multiple countries or in any other context where multiple languages are spoken, it has to be translated into all the relevant languages to be usable in a meaningful way. Translation is a very difficult process, and it can affect the comparability of data from multiple countries substantially (Schrauf, 2016, p. 86). Especially when there is no direct equivalent for a particular phrase in another language, decisions have to be made regarding how to achieve a translation that is sufficient for a cross-cultural study. Therefore, commonly there is not just one singular way to translate a particular phrase; instead, decisions are required repeatedly as to what the best options might be. At times, a literal translation may not actually be as appropriate as one that tries to genuinely reflect the meaning in another language, especially when trying to take into account the particular cultural connotations a question may have (Harkness et al., 2004).

In order to spot potential problems in translations before administering a questionnaire, a classic and often employed technique is the so-called back-translation (Su & Parham, 2002). After one expert has carried out an initial translation, another person would then be asked to translate the target-language document back into the original language. The researchers would then search for divergences in the back-translation by comparing it with the original questionnaire. Where major divergences occur, one would then analyse whether the translation was inadequate in the first place. While this approach has a high degree of plausibility at face value, it provides no guarantee that all problems will be detected (Behr & Shishido, 2016, p. 272). For example, a major error in translation may occur in the first step because of a problem with a non-existent equivalent for an idiom, but in the back-translation, the best identified alternative is very similar to the original phrase. The translation itself may be flawed and not directly comparable, but it may not actually show. Therefore, other alternatives have been suggested that can help enhance the quality of the translation. Behr and Shishido (2016, p. 271) champion the role of parallel translations. Following this approach, two translators would carry out translations independently of each other and then would jointly discuss their outcomes with each other or with a wider group of researchers to go through a reconciliation process discussing disagreements. Regardless of the precise approaches used, after the translations have been undertaken, they should ideally be examined through pretesting and piloting in every country context where they are meant to be applied.

While there are many detailed considerations regarding how best to manage cross-cultural translations and how to test their precise quality (see, e.g. Harkness et al., 2004), there are a few factors that most authors researching this field agree on. Most fundamentally, translations should always be carried out by language specialists who are able to distinguish between literal and meaning-oriented translations (Behr & Shishido, 2016); however, expertise considerations could include knowledge about both translation processes and the specific cultural and thematic contexts of the countries and topics under investigation (Su & Parham, 2002, p. 582). Expertise may be mixed in teams (Harkness et al., 2004, p. 464). It is crucial that experts can work jointly with survey developers and discuss concerns not only in a technical but also in a contextualised way. A good translation does not mean that the social concepts investigated actually have consistent meanings across countries (we will discuss this more below), but it is an essential precondition for the possibility of considering such cross-cultural comparability at all.

Post-fieldwork quality checks

We can do many things to achieve good-quality data in our surveys. However, there is still a range of issues that we should look into once we receive the data in order to assess how good the data quality ultimately is. While there are many detailed points that experienced survey developers would go through, we will focus on the three most important areas of quality checks that should be considered: (1) data cleaning, (2) sensitivity analyses and (3) assessments for cross-cultural validity.

Data cleaning

Starting to work on a new data set with data collected from survey respondents is an exciting moment, especially after lengthy periods of preparation and fieldwork.

However, there is a wide range of issues a good researcher would want to consider before diving into the data and beginning to use it, mainly to ensure that when beginning analyses, they are not negatively affected by particular errors or invalid response profiles. Errors can be caused by very trivial factors, but they can also be more systemic and require analytical techniques to identify them.

Some errors are genuinely straightforward and result from mis-labelling or mis-recording by interviewers or the people creating the data sets. Chris Welzel, Vice President of the World Values Survey Association, in an interview for this volume, described an example they encountered when receiving the data for the World Values Survey in one wave for Egypt. Examining the frequency distributions for all variables, they found that roughly 85% of people identified themselves as secular – which does not seem reasonable at all. They looked into the original coding and recordings from the interviews and found that the scale for this particular question was applied in the wrong direction compared to the values assigned in the original questionnaire. A simple coding error like this can, of course, be rectified easily; however, if not detected, it could distort findings quite dramatically. This is why survey data should always be checked – for example, by looking at the distribution of values for each variable to identify highly unlikely or extreme results that may require investigation.

Developing a good process to approach data cleaning systematically is very important in this regard (Van den Broeck et al., 2005). While the above example presents a genuine mistake, there are also survey and polling companies that have been found to manipulate and fabricate data (Kwasniewski et al., 2018) in order to reduce their costs (by having to recruit fewer people). This may manifest in many interviews apparently having been conducted at the same time, strange patterns of answer options or unreasonably perfect data, with very few missing cases (which Chris Welzel describes as an experience that led them once to find out that a survey provider they worked with was indeed creating fake data). However, detecting this after the fact can be difficult, which is why good researchers tend to always want to monitor data collection results on a continuous basis to spot irregularities during the process itself.

However, data cleaning is not only about identifying macro-level problems due to errors in coding and recording or intentional manipulation. Data cleaning is also used to identify when respondents provide careless answers to questions or do not engage with the questions at all (Meade & Craig, 2012). Especially when surveys are self-administered, researchers usually cannot discern directly from the answers whether the respondents actually read the questions properly or just randomly selected options. However, detection mechanisms during and after the data collection period can help identify such cases and exclude them from the analysis if necessary. This can be done, for example, by noting response patterns that are unlikely (e.g. always selecting the same answer option for all questions of a similar format – basically just clicking through the file). Also, the time that respondents take to answer questions is commonly recorded. If the speed is too high, meaning that it would have been impossible to actually read the questions, respondents may be screened out during the data collection itself or potentially at a later stage.

Such checks are crucial to enable researchers to have confidence that the cases they work with are genuine and that the questions have been recorded in the way they were intended.

Sensitivity analyses

While data cleaning enables us to pick up specific errors in the data, sensitivity analyses allow us to engage in more detail with our results in reference to existing research and findings. Rather than looking at the distributions within variables, we focus on looking at whether the results actually appear to be plausible given our knowledge from other studies or in direct comparison to other research. The depth and complexity of sensitivity analyses can vary substantially and depend on time resources. However, some form of checks on the results is always very helpful in appraising the degree of confidence one may have in the data. This form of external validation is also useful to explore systematically when, for example, one wants to position one's own research within a broader field explicitly. It is important to note that deviation from mainstream findings does not necessarily mean that one's own data are flawed; contextual factors may have changed or the method employed was intentionally different. Nevertheless, being able to point out such concerns is tremendously helpful both to the researcher and to the reader. When deviations are observed though, researchers should investigate what may have caused them to understand that they might be due to errors, sampling biases or genuine differences in observations. Some issues (e.g. sampling biases) can potentially be accounted for by weighting, while others require more extensive engagement. Establishing this degree of transparency either way is not just good practice but also very useful for future researchers who may wish to learn from the methods employed.

The simplest approach is to look at key results from the survey and compare them with findings in the literature on the same topic and, crucially, addressing the same context (e.g. country and age group). Simple checks of patterns in the correlations between key variables and demographic factors can be a good starting point. Often though, we aim to compare the results from a survey explicitly against some comparable external source, especially if we wish to replicate a real-life pattern. Obviously, random sampling variation needs to be kept in mind, as results will not be identical, but checking for general patterns can give a quick overview. For example, in electoral studies, researchers would commonly look at the political party choices of respondents after the data collection, considering whether the distribution is similar to that found in other polls conducted recently or, if there was an election recently, the distribution is roughly similar to actual vote choices in that election (e.g. whether the ordering of parties and their relative distances are the same). The principle is – as John Curtice outlined in his interview for this volume – that if the sampling has been done well and the weights for socio-demographic characteristics applied appropriately, the outcome for 'key variables' (in this case how somebody voted last time) should be close to the actual result.

A strong advantage of such approaches is that they are rather straightforward to explain and easy to comprehend for a reader of a study as well. However, sometimes we will want to carry out a more systematic and in-depth analysis to appraise the quality of our sample more comprehensively. One method to examine the appropriateness of a sample externally can be so-called propensity score matching techniques (or other variants thereof). The goal is to take an external source of assumed high quality (e.g. a highly regarded survey administered to the same population on a similar topic) and to examine how similar or different the distributional patterns are in a systematic fashion. A well-documented example can be found in a project examining public attitudes on constitutional change in the UK (Kenealy et al., 2017, pp. 163–172). The authors conducted a survey of adult respondents across the UK using an online web panel and a quota-sampling approach and wanted to check whether the data distributions were sufficiently good compared to a high-level face-to-face survey (the British and Scottish Social Attitudes Surveys in this case). The results show that the researchers managed to achieve a distribution in which estimates were not majorly different from those in the other survey and where deviations occurred that required more investigation, the particular implications of which could be discussed. While such detailed analysis may not always be feasible in every project undertaken, conducting some sensitivity checks and reporting their results is helpful not just for the authors but also for the researchers, subsequently aiding them in contextualising the data properly.

Cross-cultural construct validity

When conducting surveys across different countries, the primary concern in the design initially is the appropriate translation, as discussed above. In the absence of definitive and clear knowledge of the specific instruments and topic, however, we are never able to stipulate in advance whether the concepts we actually employ are going to be equivalent in their understanding and application across countries. As discussed in Chapter 4, validity is a key concern when designing survey questions. This is particularly important when we design more complex sets of questions that are *constructed* to work together to reflect a particular idea we want to operationalise and measure. Even if an adequate translation is found, it may still be that concepts differ in their meanings and social construction to an extent that could undermine the comparability of the instruments we develop for a survey. Because of these concerns, there is a substantial field of research work on cross-cultural comparability developing sophisticated methods that permit us to examine the quantitative data we obtain with regard to the equivalence of concepts across different contexts (Davidov et al., 2011).

To see how cross-national variation in concepts can be an important issue to engage with, we can initially refer to studies employing qualitative and mixed-methods approaches that enable us to illustrate what we mean when we talk about conceptual differences. A prominent example is the study of happiness and subjective well-being. Lu and Gilmour (2004) and Uchida et al. (2004), for example, present useful insights into the different conceptions of the term *happiness* held by American and Asian respondents, even when both speak English and the term is not prone to translation issues. They demonstrate that American respondents tend to conceive of happiness as a property of the individual that should be maximised through the experience of peak moments, while for many Asian respondents, happiness tends to be understood more as a property of communities that is characterised by a balance between moments of positive and negative affect. Such research indeed raises questions about the direct comparability of indicators, such as happiness scores, across different contexts. They may function, but they measure different things. This does not mean that it is a problem if the measure is understood as a very subjective notion, but concerns may be high when the measure is treated in a directly comparable manner.

Indeed, 'it could well be the case that . . . variables, although measured in a similar way across cultures, are not comparable' (Cieciuch et al., 2016, p. 630). The key concept that many researchers addressing this field focus on is *measurement invariance*. Sometimes it is also referred to as *measurement equivalence* and refers to the assessment of whether a particular construct applied to different groups (e.g. respondents in different countries) is actually measuring the same thing for each group. The focus of the analyses is to check whether the questions employed in a questionnaire behave similarly in different contexts or whether there are major variations. It is crucial to note that this does not mean that researchers are required to find the same results for a particular indicator or the same levels of a particular scale everywhere. The key question of interest is the functioning across different contexts:

> It does neither suggest that the results obtained across the various groups are identical nor that there are no differences between the groups regarding the measured construct. Instead it implies that the measurement operates similarly across the various groups to be compared and, therefore, that the results of the measurement can be meaningfully compared and interpreted as being similar across groups. (Cieciuch et al., 2016, p. 631)

Researchers who engage with questions of measurement invariance tend to use techniques that aim to understand how multiple measured items, which are understood to be part of underlying latent constructs, relate to each other and the latent construct they jointly reflect. The aim of such research is to see whether these relationships are similar or different across different country contexts, through techniques such as multi-group confirmatory factor analysis.

However, some people criticise a focus on measurement invariance to assess cross-cultural comparability. Responding to an analysis challenging key indicators measuring value change for secular and emancipative values (Alemán & Woods, 2016), Welzel and Inglehart (2016) discuss what they see as limitations of this approach. In their view, the focus on internal convergence effectively manages to establish the internal validity of concepts; however, it does not permit us to make statements about the external validity of measures. In their analyses, they show that even when assumptions of measurement invariance are not met internally, indicators may be found to behave in equivalent fashion when focussing on the relationship of those indicators with other external factors. In other words, even when technically there are internal inconsistencies, the applicability of such indicators in practice may not be affected automatically. Instead, they argue, not examining this 'external linkage' actually is a major shortcoming and should be the main focus when examining cross-cultural comparability. Essentially, they posit that external validity should be the key focus of concern, while the opposing perspective is to focus on internal validity.

This short section cannot act as an arbiter in this current debate about how best to understand and test cross-cultural comparability. But it does illustrate that the discussions we engaged in in earlier chapters regarding the quality and interpretation of survey concepts in their own right need to be considered more extensively in instances of careful cross-cultural comparison. Before undertaking the analyses to answer the main research questions, researchers should cautiously examine their concepts with regard to their applicability across different countries.

Chapter Summary

- In the previous chapters, we dealt with crucial elements of survey design: sampling, data collection mode plans and questionnaire development. All of these have to be prepared carefully to achieve good-quality outcomes. By utilising the existing knowledge from methods researchers and practitioners, we can learn how to avoid common pitfalls and reduce the risk of obtaining data that is not representative or is inadequate in measuring our concepts of interest.

(Continued)

- However, even the best planning cannot ensure that no errors would occur. As we have discussed in this chapter – partially through the exploration of real experiences of experts who develop and coordinate surveys – there are many steps which are undertaken before a survey actually goes into the field and that extend beyond sampling and questionnaire design.
- Techniques for pretesting and piloting, both qualitative and quantitative, help us to examine our question sets and our approach to running a survey. In particular, they enable us to test whether expert ideas about certain concepts actually resonate with and make sense for respondents – who ultimately will engage with these questions – in the way we intend.
- But the work on quality checks does not stop there. Even after the data collection, we should always carry out detailed checks on our data, before actually beginning the analyses for which we conducted the survey. This is always pertinent, but particularly so when we conduct comparative research. Ensuring appropriate translations before the fieldwork is a good starting point, but analysing whether our constructs empirically work across different countries and cultures is of great importance if we want to apply our measures in more than one place.
- While there are debates about how best to carry out such examinations, doing it in some way is always meaningful. Presenting the findings from such checks transparently then enables others to engage with and discuss the methods. This is useful for the particular survey project under consideration, but it also means that researchers can learn from specific projects for future work. Well-documented accounts of methods, checks and adjustments should therefore be standard and good practice for all surveys.

Further Reading

Blom, A. (2016). Survey fieldwork. In C. Wolf, D. Joye, T. Smith, & Y. Fu (Eds.), *The SAGE handbook of survey methodology* (pp. 382–396). Sage.

This is a good and concise text that discusses the different aspects that need to be considered when planning and engaging with survey fieldwork.

Davidov, E., Schmidt, P., Billiet, J., & Meuleman, B. (Eds.). (2018). *Cross-cultural analysis: Methods and applications* (2nd ed.). Routledge.

This is a book for those interested in gaining an advanced and more technical insight into cross-cultural analysis and comparability using complex survey methods.

Schrauf, R. (2016). *Mixed methods: Interviews, surveys, and cross-cultural comparisons.* Cambridge University Press.

This is a useful book for those who want to consider using a mix of different methods in cross-cultural research, connecting qualitative and quantitative approaches.

Willis, G. (2016). Questionnaire pretesting. In C. Wolf, D. Joye, T. Smith, & Y. Fu (Eds.), *The SAGE handbook of survey methodology* (pp. 359–381). Sage.

This is a concise text that introduces readers to methods for testing whether the instruments developed in a questionnaire work as they are expected to do before commencing full fieldwork.

6

CONCLUSION

This volume started with a story about well-intentioned use of survey inquiry that ended up creating a false image of what was actually going on. The message about the massive party membership surge by very young voters had not taken place, but serious media outlets reported the story and furthered the spread of incorrect information. In this case, it was a matter of incomplete communication. The press release by the Scottish Parliament committee should have clearly stated that the sample obtained was not representative or should have decided to focus not on those figures, but instead on the qualitative input generated through their engagement activities. Once 'hard' figures are out there, they are easily taken up and passed on. It is therefore all the more important that we are careful in the foundations of the design of any survey research we undertake. Any data can be analysed, but whether that data actually represents what we claim it represents is the fundamental question we need to address first. There are two things we need to do as well as possible in every single survey project: sampling and questionnaire design. We need to make sure that the cases investigated are as representative as possible of the population we plan to study, and the questions we ask our respondents need to actually operationalise the ideas and concepts we are interested in.

In providing an introduction to both of these tasks, this volume aimed to connect abstract methodological discussions to real-world practice and examples through the case studies presented and the insights from experts working in this field, both in academia and the private sector. At this stage, readers will hopefully feel that they would know not only where to start if endeavouring to begin their own survey project but also what to look out for when evaluating the work of others. Covering a wide range of topics, this volume can only act as an introduction, but it provides a solid foundation to know which areas need to be considered at the early stages of survey research. The ability to critically engage with the outputs others produce may be the most important outcome for any reader of this volume. Mistakes and misrepresentations of survey data occur on a frequent basis, but often are not intentional, of course. Being able to spot them prevents anyone engaging with such work from passing it on or using it uncritically, thus breaking chains of potentially passing on problematic or misleading information. It can be even more serious, however, when survey research is wilfully used to misinform or manipulate. While the former scenario of honest mistakes is more likely, the latter occurs, and identifying it can be challenging. As mentioned in Chapter 5, this can even occur with problematic survey companies (Kwasniewski et al., 2018), but misuse of data stretches from originators of data to its users in business, public affairs or politics. The more people are able to identify suspicious patterns and interrogate the quality of the data used, the greater are our chances of countering such abuses. Readers of this volume will hopefully join the group of people who do not just take statistics at face value but ask where the data underpinning those statistics come from and how they were created.

While achieving the best quality in our own work and being as critical as possible when evaluating data are the main goals of learning about the methods discussed in this volume, dealing with real-life insights also has an important second implication: it is not always possible to achieve the best that is theoretically possible, because practice in the real world will throw up certain barriers that we have to overcome. Chris Welzel put it very well when being interviewed for this volume: 'we have, of course, certain ideals about what we think is the role model that we try to follow. Sometimes, because of practical implications, we have to deviate a little bit from what we would consider our reference point'. This does not mean giving up on good quality or striving for representativeness. But conducting research occurs within a particular framework of constraints. Most commonly, we probably think of budgetary limits; however, it may simply be that we do not have a fully usable sampling frame or the infrastructure to utilise it properly. It could be that we could afford a certain ideal data collection method, but may not have enough time to do so, or that we find that we need to compare data over time, even when we wish we could alter the wording of a question that we had asked before but would ideally like to improve. And even when we do things 'by the book', there are still margins of errors and uncertainties that will always exist when we employ sampling techniques (as is discussed more in Volume 3 on inference in this series). John Curtice summarised this fundamental consideration well in his interview:

> Even if you're doing a probability survey, if you're trying to estimate something where the population distribution is roughly 50/50 and you get about a thousand people on a random basis, there's a 1 in 20 chance that you're going to get a number that's less than 47 or greater than 53. So sampling error is something we always live with, we know about.

That is no reason to despair. To the contrary, it allows us to evaluate the quality of our analyses. When we are able to estimate how 'good' a certain sample should be theoretically, we can identify data that appears to be suffering from problems – or data that appears to look too perfect. That is why good survey researchers and organisations do not just publish their results, but publish details on their sampling approach, the extent of aspects, such as missing cases and biases within the distribution of their sample and how they may have adjusted for them by weights. When those details are published transparently, we know how to work with the particular data in front of us and for which sort of conclusions they are more suitable and for which ones, less.

At each step of designing our sampling strategy and the questionnaire, we need to take decisions that always aim to achieve the best possible results, but under the given circumstances. As we have seen throughout the volume, there are indeed numerous decisions that we must make and that others have the right to ask us to justify. In Chapter 2, we started with conceptual questions about what the actual population for our inquiry was, but quickly moved on to the practical implications of trying to identify existing sampling frames. When those do not exist, we need to adapt and find sampling strategies that allow us to still recruit respondents meaningfully. That can happen through stratification or cluster designs that make use of certain characteristics or geography; however, it becomes increasingly more difficult the more specific our population of interest may be (e.g. a very narrow demographic or a relatively small minority group of people). Probability sampling approaches tend to be preferable, because they aim to implement or approximate randomisation procedures as closely as possible. They are not always feasible, however, so certain non-probability sampling techniques, such as quota sampling, are commonly used too. Data from such approaches can be very useful, but there are many considerations that we need to take into account in order to achieve meaningful samples. Fundamentally, regardless of which approach we choose, we need to ensure that the sort of analyses we plan to perform and the sort of conclusions we envisage to prepare can be backed up appropriately by the respective approach chosen.

But even after settling on a particular sampling approach, we are faced with the next big challenge: choosing and implementing a good data collection strategy, as we discussed in Chapter 3. While certain modes are more common for certain sampling techniques, there is no direct match. Data can be collected online from samples drawn through probability or non-probability approaches, and face-to-face surveys can be conducted with elaborate randomisation procedures or a person with a clipboard at the train station. Considerations are far-ranging, one of the most important ones being response rates, which can vary greatly for the same data collection mode by country (e.g. depending on landline penetration and phone registration records). Additionally, response can vary greatly between groups, especially when certain modes are more accessible to some (e.g. online-based systems to internet users only). Biases that can emerge because of such mode effects can be impactful and are one of the reasons why mixed-mode surveying techniques have been used increasingly by a wide range of researchers.

As we have seen in Chapter 4, the quality of our data also depends extensively on the questions we ask, not only who we engage with. A clear understanding of our research aims and questions is an elementary starting point for our design. We need to think carefully about how we can ask questions of respondents that allow us to address the core research focus – without asking directly, in particular when our core question may be complicated and wide-ranging. Being able to select the right type of question is not easy, as there are multiple implications both in terms of how people understand it and subsequently how we can analyse it. All experts interviewed for this volume agreed that it is important that questions are simple, clear and understandable, but beyond that we appreciate how people respond to what they are being presented with. We should never make the mistake of assuming that what somebody says in response to a question means that this is what they think generally, even if they had not been asked the question at all. Respondents may be happy to give a view once we ask them to think about it, but that does not mean they were thinking about the particular question all day long. It is easy to read more into a response than it reveals, so we need to remain cautious in our appraisal of how far our interpretations can go.

Finally, the discussions in Chapter 5 have shown us that planning for a good survey is crucial, but relying on our ability to prepare well would not be representing an excessive degree of confidence. Even after developing a textbook sampling strategy and coming up with carefully thought-out questions, it is hard to anticipate the unexpected. Therefore, conducting extensive checks to ensure the quality of our work can help us spot problems in advance or early or at least allow us to make adjustments later on. Piloting can take multiple forms and be undertaken to a different extent depending on resources, in particular, but checking whether the ideas one develops actually work in practice is always worth doing. This holds particularly true in new contexts, such as an area that has not been sampled before or a set of questions that was newly designed. Cognitive interviewing can be very helpful in the latter case, although it also presents challenges in its own right. No single technique is perfect, but any checks are likely to enhance the quality of our analyses. The same applies at the end of the data collection, when we want to undertake sensitivity checks to properly appraise the strengths and weaknesses of the data. All these points can become even more crucial when we conduct cross-cultural research. For such projects, we do not only need to make sure that our ideas work in practice generally, but that they work the same way in all sites of interest, so that the data is ultimately comparable.

Considering all these challenges may seem daunting at first, but it should not put anyone off from doing survey research. The efforts are worth it, because when conducted well, survey data can be very powerful and provide deep and meaningful insights for many areas of life and work. And crucially, we do not work in isolation from each other. When we present problems that occurred transparently and discuss how we engaged with them to solve those problems, we do not just help others to understand the advantages and limitations of the particular data we may have produced. We also provide insights that others can build on to enhance their own practice and ultimately help everyone conducting survey research improving the quality of outputs that may affect public and policy debates. That is why established research organisations in this field actually want their staff to engage with the debates about methodology that is happening in academia and practice. Paul Bradshaw and Susan Reid from ScotCen Social Research describe the importance of this in their own work well:

Paul Bradshaw: If we are thinking about a new project, that we haven't delivered before, we might look at what we've done that's similar in the past ... And if we haven't, who has and how did it go? ... We then draw on other reputable organisations or research teams who've delivered something like that in the past to see whether or not it's feasible and what's new and how we might expect it to go ...

Susan Reid: We're also lucky enough that we have someone whose job it is to look at those kind of things, who ... would go to conference where they're talking about other people's research, where they've looked and trialled different options and done experiments.

It is an optimistic account to finish this volume on, because it reflects on the extensive learning that occurs through survey research both in terms of methodology and ultimately the resulting findings. This also should act as a reminder that methods, like any research, evolve over time and new ideas and insights emerge constantly. New technologies or techniques are developed or tested and enable us to answer new research questions, which keeps the research exciting. Learning from the insights of the past is a great foundation for anyone engaging with social surveys to use them well, but it also means that through our work we will also contribute to the developments in the future.

GLOSSARY

Attrition: In panel studies, respondents are asked to take part in a survey at multiple consecutive time points. However, some respondents may either stop taking part at some point altogether (e.g. if they pass away) or miss participation in a wave (e.g. because they were not reachable during the fieldwork period). This 'dropout' between waves of data collection is called attrition.

Closed-ended questions: Closed questions are questions in a questionnaire that have a defined set of possible answers respondents can choose from.

Cognitive interviewing: Cognitive interviewing refers to a process through which the quality of new survey questions is analysed. Respondents are asked to answer a particular question in the same way they would in an actual survey situation, but then asked follow-up questions about what they thought about when responding to the question to better understand how their response might be interpreted and whether this matches the intention behind the question.

Fieldwork: Fieldwork refers to the process during which a survey is administered. That means the agreed framework for data collection is implemented, respondents are recruited and the questionnaire is administered to those respondents.

Open-ended questions: Open-ended questions do not present respondents with defined categories to choose from, but ask them to enter text or numbers freely to address the question asked.

Piloting: To test whether a survey works well, we often do not run fieldwork for the full sample, but only start with a small number and examine responses to identify any problems that may require adjustments before collecting data for the full sample.

Question order effect: The way people respond to a question can be impacted by the questions they were asked before that. If people were asked a set of questions about a particular issue that can make them think about the next set of questions in that particular context. Therefore, question order needs to always be considered carefully.

Sampling: Sampling is the process of selecting a group of respondents that are representative of a larger defined population through a fixed and intentional procedure.

Sampling frame: The sampling frame is the set of all possible respondents in the population of interest that we can draw our sample from.

Social desirability: Some questions may be socially or culturally rather sensitive, and respondents may be uncomfortable revealing their true views on an issue, because it may be seen as socially deviant, for example. We therefore need to carefully design questions to reduce social desirability in order to make it easier for respondents to express their actual views.

Check out the next title in the collection: *Archival and Secondary Data Analysis,* **for guidance on data archives and secondary data analysis.**

REFERENCES

Albaum, G., & Smith, S. (2012). Why people agree to participate in surveys. In L. Gideon (Ed.), *Handbook of survey methodology for the social sciences* (pp. 179–193). Springer. https://doi.org/10.1007/978-1-4614-3876-2_11

Alemán, J., & Woods, D. (2016). Value orientations from the World Values Survey: How comparable are they cross-nationally? *Comparative Political Studies, 49*(8), 1039–1067. https://doi.org/10.1177/0010414015600458

American Association for Public Opinion Research Standards Committee (2010). Research Synthesis: AAPOR Report on Online Panels. *Public Opinion Quarterly, 74*(4), 711–781, https://doi.org/10.1093/poq/nfq048

Armstrong, R. (1987). The midpoint on a five-point Likert-type scale. *Perceptual and Motor Skills, 64*(2), 359–362. https://doi.org/10.2466/pms.1987.64.2.359

Arnab, R. (2017). *Survey sampling theory and applications.* Academic Press. https://doi.org/10.1016/B978-0-12-811848-1.00002-9

Baker, R., Blumberg, S., Brick, J., Couper, M., Courtright, M., Dennis, J., Dillman, D., Frankel, M., Garland, P., Groves, R., Kennedy, C., Krosnick, J., Lavrakas, P., Lee, S., Link, M., Piekarski, L., Rao, K., Thomas, R., & Zahs, D. (2010). Research synthesis: AAPOR report on online panels. *Public Opinion Quarterly, 74*(4), 711–781. https://doi.org/10.1093/poq/nfq048

Bale, T., & Webb, P. (2015, September 2). Grunts in the ground game: UK party members in the 2015 general election [Paper presentation]. The Conference on The 2015 British General Election: Parties, Politics and the Future of the United Kingdom, UC Berkeley, CA, United States. Retrieved from https://esrcpartymembersprojectorg.files.wordpress.com/2015/09/bale-and-webb-grunts-in-the-ground-game1.pdf

Beatty, P., & Willis, G. (2007). Research synthesis: The practice of cognitive interviewing. *Public Opinion Quarterly, 71*(2), 287–311. https://doi.org/10.1093/poq/nfm006

Behr, D., & Shishido, K. (2016). The translation of measurement instruments for cross-cultural surveys. In C. Wolf, D. Joye, T. Smith, & Y. Fu (Eds.), *The SAGE handbook of survey methodology* (pp. 269–287). Sage. https://doi.org/10.4135/9781473957893.n19

Belli, R., Traugott, M., Young, M., & McGonagle, K. (1999). Reducing vote overreporting in surveys: Social desirability, memory failure, and source monitoring. *Public Opinion Quarterly, 63*(1), 90–108. https://doi.org/10.1086/297704

Billiet, J., & Loosveldt, G. (1988). Improvement of the quality of responses to factual survey questions by interviewer training. *Public Opinion Quarterly, 52*(2), 190–211. https://doi.org/10.1086/269094

Blom, A. (2016). Survey fieldwork. In C. Wolf, D. Joye, T. Smith, & Y. Fu (Eds.), *The SAGE handbook of survey methodology* (pp. 382–396). Sage. https://doi.org/10.4135/9781473957893.n25

Bowden, A., Fox-Rushby, J., Nyandieka, L., & Wanjau, J. (2002). Methods for pre-testing and piloting survey questions: Illustrations from the KENQOL survey of health-related quality of life. *Health Policy and Planning, 17*(3), 322–330. https://doi.org/10.1093/heapol/17.3.322

Brooks, L. (2015). Quarter of youngest Scottish voters have joined a party since referendum. *The Guardian.* Retrieved from www.theguardian.com/politics/2015/jan/16/quarter-under-18s-scotland-political-party-referendum

Busse, B., & Fuchs, M. (2011). The components of landline telephone survey coverage bias: The relative importance of no-phone and mobile-only populations. *Quality & Quantity, 46*(4), 1209–1225. https://doi.org/10.1007/s11135-011-9431-3

Byrne, D. (2017). What are face-to-face surveys? *Project Planner.* https://doi.org/10.4135/9781526408563

CentERdata. (2018). LISS Panel: Listening to people. Retrieved from www.lissdata.nl/

Cieciuch, J., Davidov, E., Schmidt, P., & Algesheimer, R. (2016). Assessment of cross-cultural comparability. In C. Wolf, D. Joye, T. Smith, & Y. Fu (Eds.), *The SAGE handbook of survey methodology* (pp. 630–648). Sage. https://doi.org/10.4135/9781473957893.n39

Couper, M. (2008). *Designing effective web surveys.* Cambridge University Press. https://doi.org/10.1017/CBO9780511499371

Couper, M. (2017). New developments in survey data collection. *Annual Review of Sociology, 43,* 121–145. https://doi.org/10.1146/annurev-soc-060116-053613

Cumming, R. (1990). Is probability sampling always better? A comparison of results from a quota and a probability sample survey. *Community Health Studies, 14*(2), 132–137. https://doi.org/10.1111/j.1753-6405.1990.tb00033.x

Davidov, E., Schmidt, P., & Billiet, J. (Eds.). (2011). *Cross-cultural analysis: Methods and applications*. Routledge. https://doi.org/10.4324/9780203882924

De Leeuw, E. (2005). To mix or not to mix data collection modes in surveys. *Journal of Official Statistics, 21*(2), 233–255.

De Leeuw, E., & Berzelak, N. (2016). Survey mode or survey modes? In C. Wolf, D. Joye, T. Smith, & Y. Fu (Eds.), *The SAGE handbook of survey methodology* (pp. 142–156). Sage. https://doi.org/10.4135/9781473957893.n11

De Leeuw, E., Mellenbergh, G., & Hox, J. (1996). The influence of data collection method on structural models: A comparison of a mail, a telephone, and a face-to-face survey. *Sociological Methods and Research, 24*(1), 443–472. https://doi.org/10.1177/0049124196024004002

DeMaio, T., & Rothgeb, J. (1996). Cognitive interviewing techniques: In the lab and in the field. In N. Schwarz & S. Sudman (Eds.), *Answering questions: Methodology for determining cognitive and communicative processes in survey research* (pp. 177–196). Jossey-Bass.

Dillmann, D. (2000). Mail and web-based surveys: The tailored design method. Wiley.

Dorofeev, S., & Grant, P. (2006). *Statistics for real-life sample surveys: Non-simple-random samples and weighted data*. Cambridge University Press. https://doi.org/10.1017/CBO9780511543265

Drennan, J. (2003). Cognitive interviewing: Verbal data in the design and pretesting of questionnaires. *Methodological Issues in Nursing Research, 42*(1), 57–63. https://doi.org/10.1046/j.1365-2648.2003.02579.x

Druckman, J., & Kam, C. (2011). Students as experimental participants. In J. Druckman, D. Green, J. Kuklinski, & A. Lupia (Eds.), *Cambridge handbook of experimental political science* (pp. 41–57). Cambridge University Press. https://doi.org/10.1017/CBO9780511921452.004

Eichhorn, J. (2018). Votes at 16: New insights from Scotland on enfranchisement. *Parliamentary Affairs, 71*(2), 365–391. https://doi.org/10.1093/pa/gsx037

Eichhorn, J., Paterson, L., MacInnes, J., & Rosie, M. (2014). Survey of young Scots. *AQMeN*. Retrieved from www.research.aqmen.ac.uk/scottish-independence-referendum-2014/survey-of-young-scots/

Electoral Commission. (2014). Scottish independence referendum: Report on the referendum held on 18 September 2014. Electoral Commission.

Eurostat. (2016). *Eurostat database*. Retrieved from http://ec.europa.eu/eurostat/data/database

Fitzgerald, J., Gottschalk, P., & Moffitt, R. (1998). The impact of attrition in the panel study of income dynamics on intergenerational analysis. *Journal of Human Resources, 33*(2), 300–344. https://doi.org/10.2307/146434

Fowler, F. (2012). *Applied social research methods: Survey research methods* (4th ed.). Sage.

Gardham, M. (2013, September 2). SNP poll puts Yes campaign ahead. *The Herald.* Retrieved from www.heraldscotland.com/news/13121030.SNP_poll_puts_Yes_campaign_ahead/

Garland, R. (1991). The mid-point on a rating scale: Is it desirable? *Marketing Bulletin, 2*, 66–70.

Gile, K., & Handcock, M. (2010). Respondent-driven sampling: An assessment of current methodology. *Sociological Methodology, 40*(1), 285–327. https://doi.org/10.1111/j.1467-9531.2010.01223.x

Glaser, P. (2012). Respondents cooperation: Demographic profile of survey respondents and its implication. In L. Gideon (Ed.), *Handbook of survey methodology for the social sciences* (pp. 195–207). Springer. https://doi.org/10.1007/978-1-4614-3876-2_12

Glynn, J., Kayuni, N., Banda, E., Parrott, F., Floyd, S., Francis-Chizororo, M., Nkhata, M., Tanton, C., Hemmings, J., Molesworth, A., Crampin, A., & French, N. (2011). Assessing the validity of sexual behaviour reports in a whole population survey in rural Malawi. *PLOS ONE, 6*(7), e22840. https://doi.org/10.1371/journal.pone.0022840

Groves, R.M., & Couper, M.P. (1998). *Nonresponse in household interview surveys.* John Wiley.

Groves, R., & McGonagle, K. (2001). A theory-guided interviewer training protocal regarding survey participation. *Journal of Official Statistics, 17*(2), 249–265.

Hamilton, D., & Morris, M. (2010). Consistency of self-reported sexual behvior in surveys. *Archives of Sexual Behavior, 39*(4), 842–860. https://doi.org/10.1007/s10508-009-9505-7

Harkness, J., Pennell, B., & Schoua-Glusberg, A. (2004). Survey questionnaire translation and assessment. In S. Presser, J. Rothgeb, M. Couper, & J. Lessler (Eds.), *Methods for testing and evaluating survey questionnaires* (pp. 453–473). Wiley. https://doi.org/10.1002/0471654728.ch22

Haskel, D. (2017). Direct mail response rates are at their highest point in over a decade. *IWCO Direct.* Retrieved from www.iwco.com/blog/2017/01/20/direct-mail-response-rates-and-2016-dma-report/

Heckathorn, D. (1997). Respondent-driven sampling: A new approach to the study of hidden populations. *Social Problems, 44*(2), 174–199. https://doi.org/10.2307/3096941

Henrich, J., Heine, S., & Norenzayan, A. (2010). The weirdest people in the world? *Behavioral and Brain Sciences, 33*(2–3), 61–83. https://doi.org/10.1017/S0140525X0999152X

Herbelin, T., & Baumgartner, A. (1978). Factors affecting response rates to mailed questionnaires: A quantitative analysis of the published literature. *American Sociological Review, 43*(4), 447–462. https://doi.org/10.2307/2094771

infas. (2006). *Weltwertestudie 2005/2006. Deutsche Teilstudie im Auftrag der International University Bremen (IUB)*. [World Values Study 2005/2006. German part commissioned by International University Bremen (IUB)]. infas Institut für angewandte Sozialwissenschaft GmbH.

International Foundation for Electoral Systems. (2018). *Election guide*. Retrieved from www.electionguide.org/countries/id/109/

Jann, B., & Hinz, T. (2016). Research question and design for survey research. In C. Wolf, D. Joye, T. Smith, & Y. Fu (Eds.), *The SAGE handbook of survey methodology* (pp. 105–121). Sage. https://doi.org/10.4135/9781473957893.n9

Kalton, G. (1983). *Introduction to survey sampling*. Sage. https://doi.org/10.4135/9781412984683

Kam, C., Wilking, J., & Zechmeister, E. (2007). Beyond the 'narrow data base': Another convenience sample for experimental research. *Political Behaviour, 29*(4), 415–440. https://doi.org/10.1007/s11109-007-9037-6

Kaplowitz, M., Lupi, F., Couper, M., & Thorp, L. (2012). The effect of invitation design on web survey response rates. *Social Science Computer Review, 30*(3), 339–349. https://doi.org/10.1177/0894439311419084

Karp, J., & Brockington, D. (2005). Voter turnout in five countries. *Journal of Politics, 67*(3), 825–840. https://doi.org/10.1111/j.1468-2508.2005.00341.x

Keen, R., & Apostolova, V. (2017). Membership of UK political parties (Briefing Paper No. SN05125). House of Commons Library.

Kenealy, D., Eichhorn, J., Parry, R., Paterson, L., & Remond, A. (2017). *Publics, elites and constitutional change in the UK: A missed opportunity?* Palgrave Macmillan. https://doi.org/10.1007/978-3-319-52818-2

Kennouche, S. (2015). In numbers: Scottish political party membership. *The Scotsman*. Retrieved from www.scotsman.com/news/politics/general-election/in-numbers-scottish-political-party-membership-1-3905167

Kish, L. (1962). Studies of interviewer variance for attitudinal variables. *Journal of the American Statistical Association, 57*(297), 92–115. https://doi.org/10.1080/01621459.1962.10482153

Kreuter, F. (2012). Facing the nonresponse challenge. *Annals of the American Academy of Political and Social Science, 645*(1), 23–35. https://doi.org/10.1177/0002716212456815

Krosnick, J., & Alwin, D. (1987). An evaluation of cognitive theory of response-order effects in survey measurement. *Public Opinion Quarterly, 51*(2), 201–219. https://doi.org/10.1086/269029

Kwasniewski, N., Maxwill, P., Seibt, P., & Siemens, A. (2018, February 1). Manipulation in der Marktforschung: Wie Umfragen gefälscht und Kunden betrogen werden. [Manipulation in market research: How surveys are forged and clients cheated]. *Spiegel Online*. Retrieved from www.spiegel.de/wirtschaft/unternehmen/manipulation-in-der-marktforschung-wie-umfragen-gefaelscht-werden-a-1190711.html

Lasorsa, D. (2003). Question-order effects in surveys: The case of political interest, news attention, and knowledge. *Journalism & Mass Communication Quarterly*, *80*(3), 499–512. https://doi.org/10.1177/107769900308000302

Lavallée, P., & Beaumont, J. (2016). Weighting: Principles and practicalities. In C. Wolf, D. Joye, T. Smith, & Y. Fu (Eds.), *The SAGE handbook of survey methodology* (pp. 460–476). Sage. https://doi.org/10.4135/9781473957893.n30

Lu, L. & Gilmour, R. (2004). Culture and conceptions of happiness: Individual oriented and social oriented SWB. *Journal of Happiness Studies*, *5*(3): 269–291.

McDonald, H., & Adam, S. (2003). A comparison of online and postal data collection methods in marketing research. *Marketing Intelligence & Planning*, *21*(2), 85–95. https://doi.org/10.1108/02634500310465399

McFarland, S. (1981). Effects of question order on survey responses. *Public Opinion Quarterly*, *45*(2), 208–215. https://doi.org/10.1086/268651

Meade, A., & Craig, S. (2012). Identifying careless responses in survey data. *Psychological Methods*, *17*(3), 437–455. https://doi.org/10.1037/a0028085

Mohorko, A., de Leeuw, E., & Hox, J. (2013). Coverage bias in European telephone surveys: Developments of landline and mobile phone coverage across countries and over time. *Survey Methods: Insights From the Field*. Retrieved from http://surveyinsights.org/?p=828

Moser, C., & Kalton, G. (1971). *Survey methods in social investigation* (2nd ed.). Heinemann.

Moser, C., & Kalton, G. (1992). *Survey methods in social investigation* (2nd ed.). Gower.

Nathan, G. (2001). Telesurvey methodologies for household surveys: A review and some thoughts for the future? *Survey Methodology*, *27*(1), 7–31.

Panelbase. (2013). Poll for SNP on adult residents in Scotland, 23–28 August 2013. Retrieved from www.panelbase.com/news/SNPPollTables020903.pdf

Raajmakers, Q., van Hoof, A., Hart, H., Verbogt, T., & Vollebergh, W. (2000). Adolescents' midpoint responses on Likert-type scale items: Neutral or missing values? *International Journal of Public Opinion Research*, *12*(2), 208–216. https://doi.org/10.1093/ijpor/12.2.209

Rada, V.D., & Martín, V. (2014). Random route and quota sampling: Do they offer any advantage over probably sampling methods? *Open Journal of Statistics*, *2014*, 391–401.

Salganik, M., & Heckathorn, D. (2004). Sampling and estimation in hidden populations using respondent-driven sampling. *Sociological Methodology, 34*(1), 193–239. https://doi.org/10.1111/j.0081-1750.2004.00152.x

Schonlau, M., & Couper, M. (2017). Options for conducting web surveys. *Statistical Science, 32*(2), 279–292. https://doi.org/10.1214/16-STS597

Schrauf, R. (2016). *Mixed methods: Interviews, surveys, and cross-cultural comparisons.* Cambridge University Press. https://doi.org/10.1017/9781316544914

ScotCen. (2018a). 'Moreno' national identity. *What Scotland Thinks.* Retrieved from http://whatscotlandthinks.org/questions/moreno-national-identity-5#line

ScotCen. (2018b). Should Scotland be an independent country? *What Scotland Thinks.* Retrieved from http://whatscotlandthinks.org/questions/should-scotland-be-an-independent-country#table

Scottish Government. (2015). Projected population of Scotland (2014-based). National Records of Scotland. Retrieved from www.nrscotland.gov.uk/statistics-and-data/statistics/statistics-by-theme/population/population-projections/population-projections-scotland/2014-based/list-of-tables

Scottish Parliament. (2015a). Devolution Committee: Survey of 16 and 17 year olds. *Infographic.* Retrieved from www.parliament.scot/S4_ReferendumScotlandBillCommittee/SurveyResults16-17Infographic.pdf

Scottish Parliament. (2015b). First-time voters were enthusiastic and informed, Committee survey finds. Retrieved from www.parliament.scot/newsandmediacentre/85740.aspx

Sears, D. (1986). College sophomores in the laboratory: Influences of a narrow data base on social psychology's view of human nature. *Journal of Personality and Social Psychology, 51*(3), 515–530. https://doi.org/10.1037/0022-3514.51.3.515

Squire, P. (1988). Why the 1936 Literary Digest Poll failed. *Public Opinion Quarterly, 52*(1), 125–133. https://doi.org/10.1086/269085

Stephan, F., & McCarthy, P. (1958). *Sampling opinions: An analysis of survey procedure.* Wiley.

Stoop, I. (2012). Unit non-response due to refusal. In L. Gideon (Ed.), *Handbook of survey methodology for the social sciences* (pp. 121–147). Springer. https://doi.org/10.1007/978-1-4614-3876-2_9

Stoop, I. (2016). Unit nonresponse. In C. Wolf, D. Joye, T. Smith, & Y. Fu (Eds.), *The SAGE handbook of survey methodology* (pp. 409–424). Sage. https://doi.org/10.4135/9781473957893.n27

Stoop, I., Billiet, J., Koch, A., & Fitzgerald, R. (2010). *Improving survey response: Lessons learned from the European Social Survey.* Wiley. https://doi.org/10.1002/9780470688335

Strack, F. (1992). 'Order effects' in survey research: Activation and information functions of preceding questions. In N. Schwarz & S. Sudman (Eds.), *Context effects in social and psychological research* (pp. 23–34). Springer. https://doi.org/10.1007/978-1-4612-2848-6_3

Sturgis, P., Baker, N., Callegaro, M., Fisher, S., Green, J., Jennings, W., Kuha, J., Lauderdale, B., & Smith, P. (2016). *Report of the inquiry into the 2015 British general election opinion polls*. British Polling Council and Market Research Society.

Su, C.-T., & Parham, L. (2002). Generating a valid questionnaire translation for cross-cultural use. *Americal Journal of Occupational Therapy*, *56*, 581–585. https://doi.org/10.5014/ajot.56.5.581

Sudman, S. (1976). *Applied sampling*. Academic Press.

Tourangeau, R., & Smith, T. (1996). Asking sensitive questions: The impact of data collection mode, question format, and question context. *Public Opinion Quarterly*, *60*(2), 275–304. https://doi.org/10.1086/297751

Uchida, Y., Norasakkunkit, V. & Kitayama, S. (2004). Cultural Constructions of happiness: theory and empirical evidence. *Journal of Happiness Studies 5*(3): 223–239.

Vandecasteele, L., & Debels, A. (2007). Attrition in panel data: The effectiveness of weighting. *European Sociological Review*, *23*(1), 81–97. https://doi.org/10.1093/esr/jcl021

Van den Broeck, J., Argeseanu Cunningham, S., Eeckels, R., & Herbst, K. (2005). Data cleaning: Detecting, diagnosing, and editing data abnormalities. *PLOS MEDICINE*, *2*(10), e267. https://doi.org/10.1371/journal.pmed.0020267

Welzel, C., & Inglehart, R. (2016). Misconceptions of measurement equivalence: Time for a paradigm shift. *Comparative Political Studies*, *49*(8), 1068–1094. https://doi.org/10.1177/0010414016628275

Willis, G. (2005). *Cognitive interviewing: A tool for improving questionnaire design*. Sage.

Willis, G. (2016). Questionnaire pretesting. In C. Wolf, D. Joye, T. Smith, & Y. Fu (Eds.), *The SAGE handbook of survey methodology* (pp. 359–381). Sage. https://doi.org/10.4135/9781412983655

World Values Survey. (2012). Wave 6 Official Questionnaire (Version 4). Retrieved from www.worldvaluessurvey.org/WVSDocumentationWV6.jsp

Yang, K., & Banamah, A. (2014). Quota sampling as an alternative to probability sampling? An experimental study. *Sociological Research Online*, *19*(1), 1–11. https://doi.org/10.5153/sro.3199

INDEX

Page numbers in **bold** indicate tables.

American Association for Public Opinion
    Research, 39, 40
answer option sets
    non-discrete, 66–7
    ordering effects, 72–3
    uncomprehensive, 68–9
Arnab, R., 18, 19, 20
attrition, 48–9

back-translation, 93–4
Baker, R., 39, 40
Banamah, A., 24
Beatty, P., 87, 88
Behr, D., 93, 94
Berzelak, N., 35, 39, 41–2, 43
bias
    attrition, 48–9
    interviewer effects, 90–1
    non-coverage error, 43
    non-response, 43–8, **44**, **47**
    self-selection, 3
    social desirability, 71–2
    weighting, 46–9, **47**, 50
Blom, A., 90, 91
Bradshaw, Paul, 5, 63, 83–5, 92–3, 107–8
British Polling Council, 23
British Social Attitudes Survey, 50, 97
Brockington, D., 71–2

Cieciuch, J., 98
closed-ended questions, 55–61, **61**
cluster analyses, word-based, 55
cluster sampling, 18–19, 34, 36, 38
cognitive interviewing, 87–8
comprehensive answer option sets, 68–9
computer-assisted self-administered
    interviewing, 34
confirmatory factor analyses, 76
construct validity, cross-cultural, 97–9
convenience sampling, 24–5, 33, 39

Couper, M.P., 37, 39, 40, 41, 43, 44, **44**
cross-cultural construct validity, 97–9
cross-quotas, 22, **22**, 23, 40
Cumming, R., 24
Curtice, John, 5, 27–8, 49–50, 97, 105

data cleaning, 94–6
data collection, 32–50
    attrition, 48–9
    case study, 35–6
    combining methods, 41–3
    direct mail, 37–8, 42
    face-to-face, 33–4, 42, 71, 85
    internet-based, 39–41, 42, 49, 60, 61, 90,
        92–3, 97
    non-response bias, 43–8, **44**, **47**
    random route procedure, 34, 35–6
    telephone-based, 35–7, 42
    weighting, 46–9, **47**, 50
De Leeuw, E., 35, 39, 41–2, 43
Debels, A., 48–9
Dillmann, D., 89
dimension reduction, 76
direct mail surveys, 37–8, 42
dress rehearsals, 89–90, 92

election polling
    exit polls, 27–8
    online, 49
    sensitivity analyses, 96–7
    social desirability issues, 71–2
    UK, 23, 27–8, 49–50
    US, 11
    validity and reliability, 77–8
European Community Household
    Panel, 48–9
exit polls, 27–8
expert interviews, 5, 107–8
    election polling, 27–8, 49–50, 97
    questionnaire design, 63–4

sampling narrowly defined groups, 13–15
survey quality, 83–5, 91–3, 95, 97, 105
exploratory factor analyses, 76
external reliability, 77
external validity, 77, 99
eye tracking, 90

face-to-face data collection, 33–4, 42, 71, 85

Gile, K., 26
Gilmour, R., 98
Glynn, J., 71
Groves, R.M., 44, **44**

Handcock, M., 26
happiness, conceptions of, 98
hard-to-reach groups, 25–7
Heckathorn, D., 26

Inglehart, R., 99
internal reliability, 77
internal validity, 77, 99
internet-based data collection, 39–41, 42, 49,
    60, 61, 90, 92–3, 97
interviewing
    cognitive, 87–8
    computer-assisted self-administered, 34
    face-to-face, 33–4, 42, 71, 85
    interviewer effects, 90–1
    interviewer preparation, 90–1
    piloting, 89–90
    telephone, 35–7, 42
item-non-response issues, 45–6

Kalton, G., 16, 18, 19, 20, 21, 23
Karp, J., 71–2

Lasorsa, D., 72
latent variable analyses, 76
leading questions, 70
Likert scale questions, 58–60
LISS (Longitudinal Internet studies for the
    Social Sciences) Panel, Netherlands, 41
Literary Digest, 11
Lu, L., 98

Market Research Society, 23
Martín, V., 24
measurement invariance, 98–9
Michigan Panel Study of
    Income Dynamics, 48
multiple-answer-item questions, 57–8
multistage sampling, 18–19, 34, 36, 38

non-coverage error, 43
non-discrete answer option sets, 66–7
non-discrete questions, 64–6

non-probability sampling, 15–16, 20–1
    convenience sampling, 24–5, 33, 39
    purposive sampling, 25
    respondent-driven sampling, 25–7
    snowball sampling, 14–15, 25, 26
    see also quota sampling
non-response bias, 43–8, **44**, **47**
numerical scale questions, 60–1, **61**

online data collection, 39–41, 42, 49, 60, 61,
    90, 92–3, 97
open format questions, 62
open-ended questions, 55
Ormston, Rachel, 5, 13–15, 63, 64

parallel translations, 94
piloting, 89–90
polls
    versus surveys, 27
    see also election polling
populations, 10, 12–15
probability sampling, 15–16
    cluster sampling, 18–19, 34, 36, 38
    direct mail surveys, 38
    face-to-face data collection, 33, 34
    internet-based data collection, 40–1, 42
    multistage sampling, 18–19, 34, 36, 38
    simple random sampling, 16–17
    stratified sampling, 19–20, 38
    telephone-based data collection, 35, 36–7
probing approaches to cognitive
    interviewing, 87–8
propensity score matching, 97
purposive sampling, 25

quality *see* survey quality
questionnaire design, 54–78
    case study, 73–4
    common pitfalls, 62–72
    linking questions for analyses, 74–6, **75**
    ordering effects, 72–4
    question types, 55–62, **61**
    social desirability issues, 71–2
    validity and reliability, 76–8
    see also survey quality
questions
    closed-ended, 55–61, **61**
    leading, 70
    Likert scale, 58–60
    multiple-answer-item, 57–8
    non-discrete, 64–6
    numerical scale, 60–1, **61**
    open format, 62
    open-ended, 55
    order effects, 72, 73–4
    ranking, 61
    screener, 36

sensitive, 34, 43, 44, 69, 71
single answer-item multiple choice, 55–7
quota sampling, 21–4, **22**
direct mail surveys, 38
internet-based data collection, 39–40, 41, 42
quality checks, 85, 97
telephone-based data collection, 35, 37

Rada, V., 24
random digit dialling, 36, 37
random route procedure, 34, 35–6
random sampling, simple, 16–17
ranking questions, 61
Reid, Susan, 5, 63, 83–5, 91–2, 93, 107–8
reliability, 77–8
respondent-driven sampling (RDS), 25–7

Salganik, M., 26
sampling approaches, 10
case study, 11
hard-to-reach groups, 25–7
narrowly defined groups, 13–15, 25
see also non-probability sampling;
probability sampling
sampling frames, 13–15, 17, 43
sampling modes *see* data collection
Scottish independence referendum,
2–3, 73–4, 86–7
Scottish Social Attitudes Survey, 82–5, 92, 97
screener questions, 36
self-selection bias, 3
sensitive questions, 34, 43, 44, 69, 71
sensitivity analyses, 96–7
Shishido, K., 93, 94
simple random sampling, 16–17
single answer-item multiple
choice questions, 55–7
snowball sampling, 14–15, 25, 26
social desirability issues, 71–2
Stoop, I., 44, 45

stratified sampling, 19–20, 38
Sudman, S., 12
survey design *see* questionnaire design
survey quality, 82–99, 105
case studies, 82–7
cognitive interviewing, 87–8
cross-cultural construct validity, 97–9
data cleaning, 94–6
interviewer preparation and briefings, 90–1
measurement invariance, 98–9
piloting, 89–90
propensity score matching, 97
sensitivity analyses, 96–7
translation, 93–4
surveys
defining, 27
versus polls, 27

telephone-based data collection, 35–7, 42
text mining, 55
thinking-aloud approaches to cognitive
interviewing, 87–8
translation, 93–4

Uchida, Y., 98
UK elections, 23, 27–8, 49–50
uncomprehensive answer option sets, 68–9
unit-non-response issues, 45–6
US presidential elections, 11

validity, 76–8
cross-cultural construct, 97–9
Vandecasteele, L., 48–9

weighting, 46–9, **47**, 50
Welzel, Christian, 5, 63, 95, 99, 105
Willis, G., 87, 88, 90
World Values Survey, 35–6, 74–6, **75**, 95

Yang, K., 24

www.ingramcontent.com/pod-product-compliance
Lightning Source LLC
Chambersburg PA
CBHW081746090825
30797CB00004B/20